PlaineR
Supervised and Unsupervised Algorithms Using R

Darrin Thomas

PlaineR
Supervised and Unsupervised Algorithms Using R

Darrin Thomas

SuJinSoLa
Saraburi, Thailand

Layout: Darrin Thomas
Photo Researcher: Darrin Thomas
Cover Design: Darrin Thomas

Copyright © 2018 ERT Group, publishing as SuJinSoLa, Saraburi, Thailand. All rights reserved.

SuJinSoLa

About the Author

Darrin Thomas, PhD, grew up in Sacramento, California and has over ten years of experience as a teacher and lecturer from Kindergarten to graduate school. He completed his bachelor and master degree in saxophone performance at California State University Sacramento. After working as a substitute teacher, he completed a credential in teaching at Pacific Union College. He then worked as a music teacher before moving to Thailand to work as a lecturer in Education/Psychology Department at Asia-Pacific International University (APIU). While overseas, Dr. Darrin completed his master degree in education at APIU. He then moved to the Philippines and completed his doctoral degree in education at Adventist International Institute of Advanced Studies. Currently, Dr. Darrin is a lecturer at Asia-Pacific International University. His enthusiasm for machine learning has led to works involving many different algorithms applied in an educational context. His blog is at https://educationalresearchtechniques.com

Dedication

To my Wife and Children

Contents

Preface	**v**
1 Principal Component Analysis	**1**
Chapter Objectives	1
Explaining Principal Component Analysis	1
Step 1: Data Preparation	2
Step 2: Check Normality and Dependency	3
Correlation Matrix	5
KMO Test	6
Bartlett's Test	6
Step 3: Perform Analysis	8
Step 4: Determine the Number of Components	9
Standard Deviation Method	9
Scree Plot	10
Step 5: Rotation	11
Step 6: Interpretation	15
Step 7: Using Components for Prediction	16
Conclusion	19
2 Exploratory Factor Analysis	**21**
Chapter Objectives	21
Explaining Exploratory Factor Analysis	21
Step 1: Data Preparation	22
Step 2: Check Normality and Dependency	23

	Correlation Matrix	24
	KMO Test	24
	Bartlett's Test	25
	Step 3: Determine the Number of Factors and Rotation Method	26
	Step 4: Run the Analysis	27
	Step 5: Interpretation of the Results	29
	Step 6: Predict with Factor Scores OPTIONAL	31
	Conclusion	34

3 Hierarchical Clustering — 35

- Chapter Objectives 35
- Explaining Hierarchical Clustering 35
- Step 1: Data Preparation 36
- Step 2: Determine the Number of Clusters 38
- Step 3: Cluster Analysis 40
- Step 4: Interpret the Results 42
- Conclusion 43

4 K-Means Clustering — 45

- Chapter Objectives 45
- Explaining K-Means Clustering 45
- Step 1: Data Preparation 46
- Step 2: Determine the Number of Clusters 49
- Step 4: Run Analysis and Interpret 50
- Predict with Results OPTIONAL 52
- Conclusion 53

5 Mixed Data Clustering — 55

- Chapter Objectives 55
- Explaining Mixed Data Clustering 55
- Step 1: Date Preparation 56
- Step 2: Analysis 56
- Step 3: Interpretation 57
- Conclusion 60

CONTENTS

6 Multi-Dimensional Scaling — **61**
    Chapter Objectives 61
    Explaining Multi-Dimensional Scaling 61
    Metric Data and MDS 62
        Step 1: Data Preparation 62
        Step 2: Analysis and Interpretation 62
    Non-Metric Data and MDS 66
        Step 1: Data Preparation 66
        Step 2: Analysis and Interpretation 67
    Conclusion 70

7 Market Basket Analysis — **71**
    Chapter Objectives 71
    Explaining Market Basket Analysis 71
    Step 1: Data Exploration 72
    Step 2: Analysis 75
    Conclusion 79

8 K-Nearest Neighbor — **81**
    Chapter Objectives 81
    Explaining K-Nearest Neighbor 81
    KNN for Classification 83
        Step 1: Data Preparation 83
        Step 2: Model Development 85
        Step 3: Model Testing 87
    KNN for Regression 90
        Step 1: Data Preparation 90
        Step 2: Model Development 92
        Step 3: Model Testing 93
    Conclusion 94

Preface

The rationale behind this text is to provide the reader with easy to understand examples of various algorithms commonly used in the machine learning data science field. The goal was not to provide extensive information on the details of statistics but rather to provide a text that explains how to run and interpret the code of an analysis.

My own background in the use of machine learning algorithms come from my studies and research as a lecturer. I have used these algorithms and taught others how to for many years and this text is a result of the experience gained from helping others in this field.

This book is focused primarily on unsupervised learning algorithms with a brief look at supervised learning algorithms as well. The first seven chapters focus on unsupervised learning algorithms focusing on such topics as clustering, components, and association rules. The last chapter looks at supervised learning with an explanation of K Nearest Neighbor.

In this text, we assume that you are already familiar with statistics and have a working knowledge of R. This is not a introductory text but design for people familiar with the field but looking to fill gaps in their knowledge about it

Hopefully, this book will provide you with the tools you need to achieve the goals you have in relation to statistical analysis.

Chapter 1

Principal Component Analysis

Chapter Objectives

- To explain the characteristics of principal component analysis.
- To explain the steps involved in conducting a principal component analysis.

Explaining Principal Component Analysis

The primary purpose of principal component analysis (PCA) is dimension reduction. By the word dimension we mean variable. There are at least two reasons for this

- When you have a large number of variables
- The variables are highly correlated

A large number of variables is unwieldy and hard to interpret. In addition, high correlation among variables will skew results. PCA will address one or both of these issues while also removing redundancy without a lost of too much information.

PCA achieves its goals in particular by transforming the data. The word transform in algebra often means to move something that it plotted on a set of axis. In PCA we transform or move the axes in order to better understand the patterns in the data. Generally, the transformation is orthogonal or at a right angle although there are other choices. The mathematical explanation of this is beyond this book. Instead we are focusing on how to do this rather than why it works.

One technical point to make is that PCA relies on to broad groups of algorithms. These are r-mode and q-mode. R-mode focuses on eigen decomposition of the correlation matrix while q-mode focuses on the singular value decomposition. If this is confusing please do not worry about it. For our purposes we will use q-mode.

For PCA, there are 6 main steps with an optional seventh step. They are as follows

1. Data Preparation

2. Check normality and dependency

3. Run analysis

4. Determine the number of components to extract

5. Select rotation method

6. Interpret components

7. Use factors score for additional analysis (OPTIONAL)

Step 1: Data Preparation

We will go through each of these steps using the `Carseats` dataset from the `ISLR` package. Below is some initial code

CHAPTER 1. PRINCIPAL COMPONENT ANALYSIS

```
library(ISLR)
data("Carseats")
```

One thing we have to do before the analysis is remove any categorical variables. PCA only works with continuous variables. If you look at the data using the `View` function you will see the following categorical variables.

- ShelveLoc

- Urban

- US

To understand what these variables are you can use `?Carseats` command to learn more. The code below will remove the unwanted variables.

```
Carseats1<-Carseats[,-c(7,10,11)]
```
This code simply removes variables 7, 10 and 11 which are the categorical variables. The new name for our data set is `Carseats1`. We will now continue our analysis.

Step 2: Check Normality and Dependency

Step involves testing assumptions of PCA. Univariate and multivariate normality can be assessed using the `mvn` function from the `MVN` package. Below is the code for the analysis followed by the output.

```
> mvn(Carseats1)
$multivariateNormality
             Test         Statistic                  p value Result
1 Mardia Skewness    131.5311622324      0.222296307400516    YES
2 Mardia Kurtosis  -4.16529518764276  3.10949968922447e-05     NO
3             MVN              <NA>                   <NA>     NO

$univariateNormality
          Test    Variable Statistic  p value Normality
1 Shapiro-Wilk       Sales    0.9952    0.254       YES
2 Shapiro-Wilk    CompPrice    0.9984   0.9772      YES
3 Shapiro-Wilk      Income    0.9611   <0.001        NO
4 Shapiro-Wilk Advertising    0.8735   <0.001        NO
5 Shapiro-Wilk  Population    0.9520   <0.001        NO
6 Shapiro-Wilk       Price    0.9959   0.3902       YES
7 Shapiro-Wilk         Age    0.9567   <0.001        NO
8 Shapiro-Wilk   Education    0.9242   <0.001        NO

$Descriptives
              n       Mean     Std.Dev Median Min    Max    25th    75th        Skew    Kurtosis
Sales       400   7.496325    2.824115   7.49   0  16.27    5.39    9.32  0.18417098 -0.10934021
CompPrice   400 124.975000   15.334512 125.00  77 175.00  115.00  135.00 -0.04243445  0.01107279
Income      400  68.657500   27.986037  69.00  21 120.00   42.75   91.00  0.04907427 -1.09628725
Advertising 400   6.635000    6.650364   5.00   0  29.00    0.00   12.00  0.63479687 -0.56550852
Population  400 264.840000  147.376436 272.00  10 509.00  139.00  398.50 -0.05084308 -1.21128171
Price       400 115.795000   23.676664 117.00  24 191.00  100.00  131.00 -0.12434811  0.41415856
Age         400  53.322500   16.200297  54.50  25  80.00   39.75   66.00 -0.07660383 -1.14453644
Education   400  13.900000    2.620528  14.00  10  18.00   12.00   16.00  0.04367733 -1.30562560
```

As you can see there are three tables

- multivariate normality
- univariate normality
- descriptive stats

We are only going to interpret the first two tables. Descriptive should be self-explanatory by this point in your statistical life.

The `multivariateNormality` table tells if your variables are normal for skewness and kurtosis. It appears that our data passes for skewness but fails for kurtosis. The third row is not applicable to our analysis.

The `univariateNormality` table looks at each variable individual. Several of our variables pass and several do not. Normality is not critical for PCA however it can have implications when conducting factor analysis, which is the subject of chapter 2. For our purposes we are done with this step.

Next, we will assess the dependency of the variables. This will be done with three forms of analysis.

- Correlation matrix
- KMO analysis
- Bartlett's test

Correlation Matrix

The correlation matrix provides an indication of the strength of the association between the variables. Generally, we want to see strong correlations when considering the use of PCA. Below is the code and output for the correlation matrix.

```
> library(ellipse)
> plotcorr(cor(Carseats1))
```

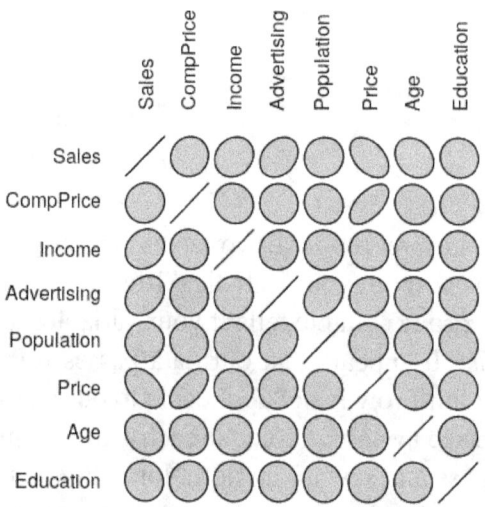

In simple terms, the bigger the circle the weaker the association and the smaller the circle the stronger the association. You can see from the output that there are few strong associations in the dataset. This would be a sign that we should not consider using PCA for further analysis. However, the purpose here is to demonstrate PCA, as such the actual results are not important.

KMO Test

The Kaiser-Meyer-Olin (KMO) index is a number between 0 and 1. This index provides an insight into the the strength of the partial correlations of the variables. If this does not make sense keep in mind that the closer the KMO is to 1 the better potential for the use of PCA.

To do this analysis you will need the psych package. The function that is used is the KMO function. Below is the code and output for the KMO analysis.

```
> library(psych)
> KMO(Carseats1)
Kaiser-Meyer-Olkin factor adequacy
Call: KMO(r = Carseats1)
Overall MSA =  0.3
MSA for each item =
    Sales  CompPrice    Income  Advertising  Population   Price    Age  Education
     0.25      0.30       0.35         0.28        0.54    0.32   0.23       0.51
```

Generally, anything below 0.5 is awful. The overall MSA is 0.3 and most of the variables are 0.3 and lower. This is another indicator that are dataset is not appropriate for PCA. However, for the purpose of demonstration we will continue.

Bartlett's Test

Bartlett's test determines if there are statistical significant relationships between the variables. This is done by comparing the correlation matrix with the identity matrix. To put it simply this test should be significant which means there is a difference between the two matrices. Remember that we want variables with strong associations when considering the use of PCA.

For this code we need the `cortest.Bartlett` function from the `psych` package. Below is the code and output for Bartlett's test.

```
> cortest.bartlett(Carseats1)
R was not square, finding R from data
$chisq
[1] 525.7253

$p.value
[1] 3.344469e-93

$df
[1] 28
```

Are results are significant which means that according to Bartlett's test PCA is appropriate. This is why it is important to run multiple test as each test gives you a slightly different conclusion that helps you to make a decision about how to proceed with your analysis.

We have already done a lot of preliminary analysis and it is easy to forget what has happen. The table below summaries the results for step 1 of our analysis.

Test	Result
Multivariate normality	
Mardia Skewness	FAILED
Mardia Kurtosis	FAILED
Dependency	
Correlation Matrix	WEAK
KMO Analysis	FAILED
Bartlett's Test	PASSED

You can make your own conclusion about what to do but we will now move to the next step and that is performing the actual analysis of the data with PCA.

Step 3: Perform Analysis

We can now begin our analysis. We will use the `prcomp` function to do this. We need to use an argument within the function called `scale` in order to scale the data. Scaling the data involves subtracting the mean and dividing by the standard deviation of each example in the dataset. Doing so allows all the data to be set to the same standard of measurement.

This is important because PCA is sensitive to scale in which larger numbers will have a stronger influence on the development of the components. Below is the code and output for our PCA.

```
> prcomp(Carseats1,scale=T)
Standard deviations (1, .., p=8):
[1] 1.3347648 1.1953039 1.0812374 1.0039170 0.9784191 0.9182958 0.8011481 0.4126966

Rotation (n x k) = (8 x 8):
                    PC1         PC2         PC3         PC4         PC5         PC6         PC7         PC8
Sales        -0.459547861  0.3999673 -0.4124553  0.07432373  0.01344899  0.3476629 -0.26080735  0.5139335673
CompPrice     0.493402225  0.4265150 -0.2154280 -0.05538185  0.09017909  0.3374057 -0.40296098 -0.4907244143
Income       -0.209069439  0.1104214 -0.1490295 -0.66610835  0.54753417 -0.4054273 -0.08697526 -0.0803571643
Advertising  -0.221388474  0.4944362  0.3449768  0.19596898  0.37178268  0.2224361  0.56395059 -0.2082466891
Population   -0.187147357  0.3304551  0.5977697  0.17522708 -0.12588142 -0.3706857 -0.56079207 -0.0006552257
Price         0.637433459  0.2916113  0.1589820 -0.11384853  0.11392389 -0.1273998  0.15088707  0.6468604187
Age           0.003327009 -0.4271835  0.4306838 -0.12047650  0.44419256  0.5529864 -0.29682063  0.1621623165
Education     0.106256785 -0.1572500 -0.2723239  0.67165769  0.57231024 -0.3033254 -0.13652788  0.0226369766
```

The information in the printout is the correlation that each variable has with the components before we rotate the axis in a future step. You can see that we have 8 components because we also have 8 variables. At the top of the printout are the standard deviations which shows how much of the information was preserved by each component. Values above one are components that explain more than the original variables.

The printout we have from R right now is hard to understand. We can improve this by using the `summary` function on the code we used previously. Below is the code followed by the results.

```
> summary(prcomp(Carseats1,scale=T))
Importance of components:
                          PC1    PC2    PC3    PC4    PC5    PC6     PC7     PC8
Standard deviation     1.3348 1.1953 1.0812 1.0039 0.9784 0.9183 0.80115 0.41270
Proportion of Variance 0.2227 0.1786 0.1461 0.1260 0.1197 0.1054 0.08023 0.02129
Cumulative Proportion  0.2227 0.4013 0.5474 0.6734 0.7931 0.8985 0.97871 1.00000
```

The printout above is much clearer. We have the standard deviation of each component, the proportion of variance explained, and the cumulative proportion of the variance explained. Notice that the components are arranged in descending order based on the standard deviation/proportion explained. In addition, if you add up all the components variance you get to 100% as you can see in the table.

The general rule is that if the standard deviation is greater than 1 you may want to keep the component. With this understanding it appears that we may want to keep 4 components. In the next step, we will look into additional ways to determine the number of components to keep.

Step 4: Determine the Number of Components

We will look at two ways to determine the number of components to keep from a PCA they are as follows...

1. Standard deviations greater than 1

2. Scree plot

Before going forward it is important to note that the number of components to keep is highly subjective and should be based on industry knowledge as well as statistical output. There is no single way to do this the purpose of these analysis tools is to guide your thinking but not to do the thinking for you/

Standard Deviation Method

We have already talk about this option. This method suggest keeping all components that have a standard deviation greater than one. Let's look at our output a second time.

```
> summary(prcomp(Carseats1,scale=T))
Importance of components:
                          PC1    PC2    PC3    PC4    PC5    PC6     PC7     PC8
Standard deviation     1.3348 1.1953 1.0812 1.0039 0.9784 0.9183 0.80115 0.41270
Proportion of Variance 0.2227 0.1786 0.1461 0.1260 0.1197 0.1054 0.08023 0.02129
Cumulative Proportion  0.2227 0.4013 0.5474 0.6734 0.7931 0.8985 0.97871 1.00000
```

The output recommends 4 components. If we keep 4 components we would still have 67% of the variance explained. The question we have to ask is whether this is enough. In some disciplines the answer is yes while in others the answer is no. For now, we are going to state that the standard deviation method recommends 4 components. The scree plot method below will provide another perspective on how to determine the number of components.

Scree Plot

The scree plot is simply a visual of what we did with the standard deviation method. To do this analysis, you will need the psych package and you will use the VSS.scree function along with the cor function on our dataset. For this analysis, the data does not need to be scaled. The code and output is below.

```
> library(psych)
> VSS.scree(cor(Carseats1))
```

This analysis indicates that perhaps we only need 3 components. This is because the 4th component is not quite above the cutoff line completely.

CHAPTER 1. PRINCIPAL COMPONENT ANALYSIS

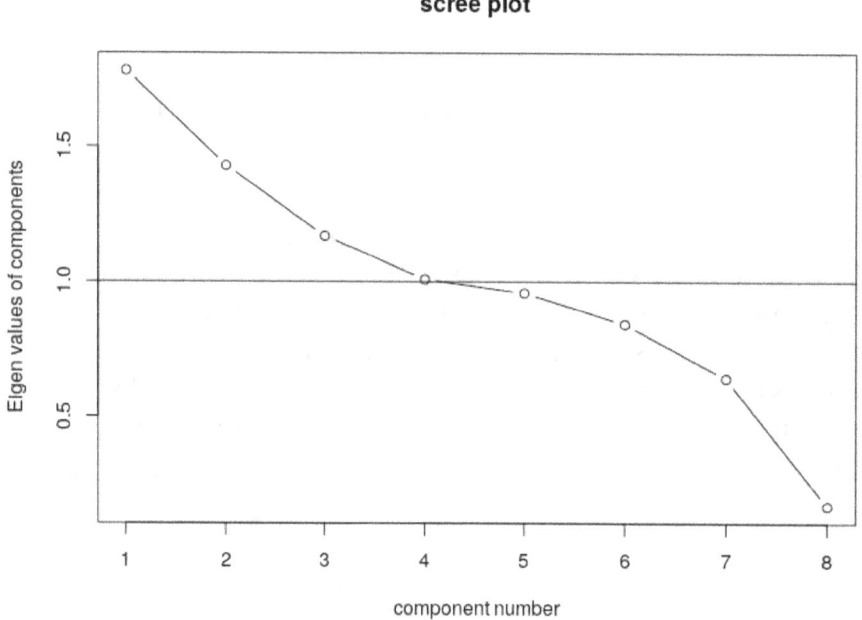

With this analysis we can decide if we want to use 3 or 4 components going forward. Normally, simpler is better. Therefore, in this example we will use 3 components going forward. This reduces the amount of variables we have to work with by 5 (8 - 3 = 5).

We now need to decide on how to rotate the data in order to enhance our ability to interpret it. This will be our next step.

Step 5: Rotation

Rotating the axes helps us to interpret the loadings of each variable on each component. There are two umbrella types of rotation and each of these umbrella types have several methods of rotation under them. The two types, as well as common methods, are described below.

- Orthogonal: New axes is perpendicular (orthogonal) to the original.

Therefore, there is no correlation between the components. Examples include...

- Varimax
- Quartimax
- Equimax

- Oblique: The new axes are not necessarily orthogonal to each other. Therefore, there may be some correlation between the components. Examples include...

  - Promax
  - Oblimin

We are not going to use all of these methods. Instead, we will use varimax from the orthogonal family and promax from the oblique family for the purpose of comparison.

To do the rotations we need to rerun are data but this time save it as an object. We will then feed the object to the function for the rotation.

In the code below we make an object called `pcResults` which has all of our PCA information. We then use the `varimax` function and give it our object `pcResults`. However, we only want `varimax` to access the `rotation` table in our object. In addition, we only want it to access the first three components of the `rotation` table. The code is below with the output.

CHAPTER 1. PRINCIPAL COMPONENT ANALYSIS

```
> pcResults<-prcomp(Carseats1,scale=T)
> varimax(pcResults$rotation[,1:3])
$loadings

Loadings:
              PC1     PC2     PC3
Sales                 0.732
CompPrice     0.675   0.104
Income                0.269
Advertising           0.178   0.612
Population                    0.702
Price         0.653  -0.285
Age          -0.318  -0.510
Education                    -0.330

                PC1   PC2   PC3
SS loadings    1.000 1.000 1.000
Proportion Var 0.125 0.125 0.125
Cumulative Var 0.125 0.250 0.375

$rotmat
          [,1]       [,2]       [,3]
[1,]  0.7832706 -0.5442945 -0.3003842
[2,]  0.6047017  0.5548860  0.5713470
[3,] -0.1443021 -0.6291621  0.7637617
```

The table begins with the loadings. This tells you the correlation that the variable has with the component. Low values (below 0.1) are left out in order to add interpretation. For example, CompPrice has a strong correlation with component 1 but a weak one with component 2. Overall, three variables load well on component 1 (CompPrice, Price, Age).

It is important to note that these loadings are the rotated results of the correlations that we saw on page 9. You can now see how transformation of the axis can help to better interpret the components. For example, Sales used to have a negative correlation with component 1 and now has a positive correlation.

The next table is the variance explained. If you carefully you will see that these numbers are different from our unrotated results. This is because the rotation changes the influence of the difference variables and thus the proportion explained. The three components now explained 38% of the variance. Before these three components explained 54% of the variance.

The last table shows the rotation matrix which indicates the standardized scores. It is the orthogonal matrix used for creating the new loadings from the unrotated ones.

We will now do the oblique rotation. To do this we will need the `GPArotation` package and we will use the `promax` function for the promax rotation. The rest of the code is the same.

```
> library(GPArotation)
> promax(pcResults$rotation[,1:3])
$loadings

Loadings:
              PC1    PC2    PC3
Sales       -0.128  0.737
CompPrice    0.665  0.102
Income      -0.101  0.271
Advertising         0.201  0.616
Population                 0.701
Price        0.681 -0.283
Age         -0.269 -0.509
Education                 -0.330

                 PC1   PC2   PC3
SS loadings    1.009 1.011 1.003
Proportion Var 0.126 0.126 0.125
Cumulative Var 0.126 0.253 0.378

$rotmat
             [,1]        [,2]        [,3]
[1,]  0.83254016 -0.5576072 -0.2991686
[2,]  0.55694538  0.5790443  0.5832255
[3,] -0.07767362 -0.6042008  0.7569698
```

You can see that the numbers are about the same. Selecting a rotation method is domain specific as well as left to personal preference.

We will now proceed to interpreting our analysis. For this we will use the results from our varimax rotation and will try to explain what the components mean.

Step 6: Interpretation

Interpreting the components is subjective and everyone will see things differently. To address this, you have to keep in mind what exactly it is you want to know, your expertise, and the overall simplicity of your interpretation. With this below are the varimax results again.

```
> pcResults<-prcomp(Carseats1,scale=T)
> varimax(pcResults$rotation[,1:3])
$loadings

Loadings:
            PC1    PC2    PC3
Sales              0.732
CompPrice   0.675  0.104
Income             0.269
Advertising        0.178  0.612
Population                0.702
Price       0.653 -0.285
Age        -0.318 -0.510
Education                -0.330

                PC1   PC2   PC3
SS loadings    1.000 1.000 1.000
Proportion Var 0.125 0.125 0.125
Cumulative Var 0.125 0.250 0.375

$rotmat
            [,1]       [,2]       [,3]
[1,]   0.7832706 -0.5442945 -0.3003842
[2,]   0.6047017  0.5548860  0.5713470
[3,]  -0.1443021 -0.6291621  0.7637617
```

The first component is strongly related to pricing so we can call this component "pricing" if we wanted to give it a name. Component 2 is tricky. It is focused on sales and age. It is difficult to explain this without additional domain knowledge. Component 3 is our "mass media reach" component as it is comprised mostly of advertising and the population size.

Step 7 is using the components from our PCA for prediction. This is an optional step that sometimes is done.

Step 7: Using Components for Prediction

We are going to make a logistic regression model using our three components to predict the location of a store in our `Carseats` dataset. The location variable was one of the categorical variables we did not include in our PCA. To complete this task there are several steps we need to perform as outlined below.

A. Make a dataset with the component scores

B. Add the dependent variable to the dataset with the component scores.

C. Create our train and test datasets

D. Create our model

E. Predict with our model

F. Test our model

The first step is to make a dataset that has the component scores. The component scores provides information on how well an individual example loads to each component. To do this we have to access information inside the `pcResults` object. We save the information in a new object called `pcreg`. Below is the code.

```
> pcreg<-as.data.frame(pcResults$x)
> head(pcreg)
       PC1        PC2        PC3         PC4        PC5        PC6        PC7         PC8
1  0.1377082  1.1770633 -0.8219688  0.90275886  0.7923956 -0.1796662 -0.1403810 -0.17306408
2 -2.2455352  0.2647584  0.7241841 -0.01838647 -0.6302855  1.8020698  0.6803588  0.07816016
3 -1.7079030 -0.3198451  0.2725008  0.65883526 -0.9636569  1.2481187  0.2188984 -0.09651509
4 -1.1444072 -0.1383694  0.5464369 -0.45467111  0.2241154 -1.0886632 -1.0235610 -0.26653173
5  1.4094503  0.4600585  0.1728241 -0.22825229 -0.8381810 -0.7870257 -0.2864344 -0.82262499
6 -2.5024706  0.3037613  0.7254173  0.06897298  1.9562318  0.1979208 -1.6177275 -0.62480468
```

CHAPTER 1. PRINCIPAL COMPONENT ANALYSIS

We now need to add the dependent variable of this analysis to our new dataset `pcreg`. This will come from the `Carseat` dataset. Below is the code.

```
pcreg$US<-Carseats$US
```

We can now make our train and test sets. We develop our model with the train data and then we test it with the test set. This means we need to split the data into two portions one for training and the other for testing. We will do a 70/30 split for this. The coding below is somewhat complicated but what it does is divide the dataset into the two groups that we need. The purpose of setting a seed is to make sure you always get the exact same results. The code is below.

```
set.seed(234)
ind<-sample(2,nrow(pcreg),replace=T,prob = c(0.7,0.3))
train<-pcreg[ind==1,]
test<-pcreg[ind==2,]
```

We can now develop our logistic regression model with the code below.

```
model <- glm(US PC1+PC2+PC3,family=binomial(link='logit'),
data=train)
```

We are not going to explain how to interpret logistic regression as I cover that in a different book. Instead, we are going to move predicting. To do this we need to use the `predict` function. Inside this function we will put our model and indicate that we want the response probabilities. The results of this we will save inside our `train` dataset in a variable called `probs`. The code is below.

```
train$probs<-predict(model, type = 'response')
```

We now need to create a variable inside our object called `predict`. At first, this variable will only contain 306 "no"s. Why 306? Because this is how many examples there are in my `train` dataset and you can check this for yourself using the `str` function. Here is the code for making the initial `predict` variable.

```
train$predict<-rep('No',306)
```

Now we will replace some of the values in the `predict` variable based on the probabilities in the `probs` variable. Simply in probability greater than 0.5 in `probs` should be coded as "yes" in our `predict` variable. The code is below.

```
train$predict[train$probs>0.5]<-"Yes"
```

We are now ready to see how accurate our model is. We will compare the predicted values with the actual values. The code is below.

```
> table(train$predict,train$US)

      No Yes
  No  61  39
  Yes 51 155
> mean(train$predict==train$US)
[1] 0.7058824
```

Our model using three components has an accuracy of 71%. Whether this is good or bad is subjective. We will now test our model with the `test` set. The code is mostly a repeat. We do need to indicate new data in the `predict` function but everything else should look familiar. The code is below.

```
> test$prob<-predict(model,newdata = test, type = 'response')
> test$predict<-rep('No',94) #sample size of test set
> test$predict[test$prob>0.5]<-"Yes"
> table(test$predict,test$US)

      No Yes
  No  12   6
  Yes 18  58
> mean(test$predict==test$US)
[1] 0.7446809
```

Our results stayed about the same with the accuracy improving to 74%. This means that are model may perform well on other datasets.

Conclusion

This chapter provided an overview of the use of PCA. We learned how to setup, run, and interpret the results of a PCA. PCA is one of the more foundational examples of unsupervised machine learning which does not involve a dependent variable.

Chapter 2

Exploratory Factor Analysis

Chapter Objectives

- To explain the characteristics of factor analysis.
- To explain the steps involved in conducting a factor analysis.

Explaining Exploratory Factor Analysis

The goal of factor analysis (FA) is the same as PCA but with a slightly different emphases. With FA you are trying to uncover the hidden or latent variables to better understand the internal structure of the data. PCA is often used for additional computation while FA is used for explaining the dataset. Having said this is it does not mean that some sort of additional prediction is never used with FA as this does happen.

In addition, the steps for FA are slightly different from PCA. Below are the steps

1. Data preparation
2. Check normality and dependency
3. Determine the number of factors and the rotation method

4. Run analysis

5. Interpret

6. Use for prediction OPTIONAL

Step 1: Data Preparation

We will perform these steps using the `Hitters` dataset from the `ISLR` package. This dataset has baseball statistics and salary of major league baseball players. Below is the initial code.

```
library(ISLR)
data("Hitters")
```

We need to do some additional data preparation. There is a lot of missing data in the `Salary` variable in our dataset. You can check for yourself if you desire. We need to remove this and this will be done with the `na.omit` function in the code below.

```
hitters<-na.omit(Hitters)
```

We now need to remove the categorical variables as factor analysis can only be used for continuous variables. In addition, we need to remove the continuous variable `Salary` because we want to predict salary a the end of the chapter. To achieve this we subset the variables we want using brackets and the c function, which only takes the variables that we want. The numbers in the brackets is the position of the variable in the dataset. Below is the code.

```
hit<-hitters[,c(1:13,16:18)]
```

The final name of our dataset going forward is `hit`. Our dataset is ready for analysis. Most of these steps we did in the prior chapter so there will be less explanation.

Step 2: Check Normality and Dependency

We will use the mvn function from the MVN package. The code and output are below. The descriptive stats are not print due to space constraints.

```
> mvn(hit)
$multivariateNormality
            Test         Statistic p value Result
1 Mardia Skewness 5143.37049245713       0     NO
2 Mardia Kurtosis 47.5715148861718       0     NO
3             MVN              <NA>   <NA>     NO


$univariateNormality
           Test    Variable Statistic  p value Normality
1  Shapiro-Wilk     AtBat      0.9690  <0.001     NO
2  Shapiro-Wilk     Hits       0.9832  0.0034    NO
3  Shapiro-Wilk     HmRun      0.9256  <0.001     NO
4  Shapiro-Wilk     Runs       0.9771  3e-04     NO
5  Shapiro-Wilk     RBI        0.9619  <0.001     NO
6  Shapiro-Wilk     Walks      0.9647  <0.001     NO
7  Shapiro-Wilk     Years      0.9257  <0.001     NO
8  Shapiro-Wilk     CAtBat     0.8761  <0.001     NO
9  Shapiro-Wilk     CHits      0.8628  <0.001     NO
10 Shapiro-Wilk     CHmRun     0.7540  <0.001     NO
11 Shapiro-Wilk     CRuns      0.8511  <0.001     NO
12 Shapiro-Wilk     CRBI       0.8268  <0.001     NO
13 Shapiro-Wilk     CWalks     0.8046  <0.001     NO
14 Shapiro-Wilk     PutOuts    0.7577  <0.001     NO
15 Shapiro-Wilk     Assists    0.7854  <0.001     NO
16 Shapiro-Wilk     Errors     0.9172  <0.001     NO
```

It appears that we failed all the test. Our data has no signs whatsoever of normality. This is critical for EFA and would mean that we would probably stop the analysis if this was for a real application.

We will now deal with dependency. We will use the same three forms of analysis as in the prior chapter.

- Correlation matrix

- KMO analysis

- Bartlett's test

Correlation Matrix

The code and output for the correlation matrix is below

```
> library(ellipse)
> plotcorr(cor(hit))
```

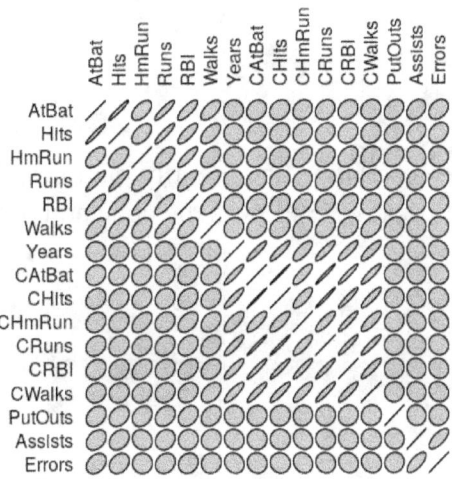

You can see from the output that there are several strong associations in the dataset. This a sign that EFA may be appropriate.

KMO Test

We will now do the KMO test. You will need the `psych` package.

CHAPTER 2. EXPLORATORY FACTOR ANALYSIS

```
> KMO(hit)
Kaiser-Meyer-Olkin factor adequacy
Call: KMO(r = hit)
Overall MSA =  0.72
MSA for each item =
   AtBat    Hits   HmRun    Runs     RBI   Walks   Years  CAtBat   CHits  CHmRun   CRuns    CRBI  CWalks
    0.78    0.64    0.63    0.69    0.77    0.70    0.94    0.78    0.64    0.67    0.69    0.72    0.75
 PutOuts Assists  Errors
    0.85    0.60    0.63
```

These results also look promising. The values are close to 1 indicating strong partial correlations. This is another sign that EFA may be appropriate.

Bartlett's Test

Bartlett's test compares the correlation and identity matrices. This test is also available in the `psych` package. Below is the code and output.

```
> cortest.bartlett(hit)
R was not square, finding R from data
$chisq
[1] 7720.791

$p.value
[1] 0

$df
[1] 120
```

More good news. Bartlett's test is significant indicating that EFA is appropriate.

The table below summaries the results for step 1 of our analysis.

Test	Result
Multivariate normality	
Mardia Skewness	FAILED
Mardia Kurtosis	FAILED
Dependency	
Correlation Matrix	MODERATE
KMO Analysis	PASSED
Bartlett's Test	PASSED

The results were great for dependency but bad for normality. This is were you have to decide if this can be ignored or if it is too much of a concern. Since are purpose is learning and instruction we will continue.

One thing that needs to be mentioned is scaling. For PCA, you must scale the data for accurate results. However, for EFA this is not necessary. Therefore, we will not scale our data before we continue.

Step 3: Determine the Number of Factors and Rotation Method

At step 3, there is a departure in how PCA and EFA are performed. In PCA, you run the analysis at this step and determine the number of components and rotation method later. In EFA, generally it is better to determine the number of factors and rotation method before analysis.

In order to determine the number of factors we will do a parallel analysis. The parallel analysis creates a random dataset that generates a correlation matrix. This is used to determine the number of factors to consider. For the rotation method we will simply use varimax as it is a commonly used method.

To do the parallel analysis we will use the `fa.parallel` function from the `psych` package. Below is the code and output for the parallel analysis.

```
> library(psych)
> fa.parallel(hit)
The estimated weights for the factor scores are probably incorrect.  Try a different factor extraction method.
Parallel analysis suggests that the number of factors =  3  and the number of components =  3
```

CHAPTER 2. EXPLORATORY FACTOR ANALYSIS

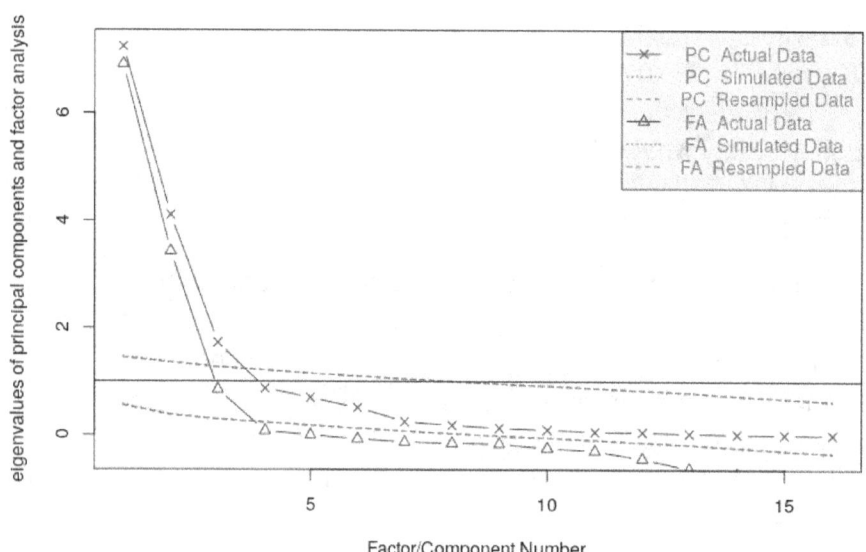

you can see that we get a warning message about the factor scores. This may be because of the problems we had with multivariate normality. If you look at the plot it has the results for both PCA and EFA as well as for the actual data and resampled data. The parallel analysis recommends three factors and we will go with this recommendation moving forward to step 3.

Step 4: Run the Analysis

To do the analysis we will use the `fa` function from the `psych` package. We will set the number of factors and the rotation inside the function. The code and a portion of the output are below.

```
faResults<-fa(hit,nfac=3,rotate='varimax')
faResults
```

```
> faResults<-fa(hit,nfac=3,rotate='varimax')
The estimated weights for the factor scores are probably incorrect.  Try a different factor extraction method.
> faResults
Factor Analysis using method =  minres
Call: fa(r = hit, nfactors = 3, rotate = "varimax")
Standardized loadings (pattern matrix) based upon correlation matrix
         MR1   MR2   MR3   h2    u2    com
AtBat    0.08  0.90  0.34  0.93  0.066 1.3
Hits     0.08  0.88  0.31  0.88  0.120 1.3
HmRun    0.16  0.74 -0.24  0.64  0.364 1.3
Runs     0.06  0.92  0.16  0.88  0.124 1.1
RBI      0.19  0.90 -0.02  0.85  0.153 1.1
Walks    0.23  0.64  0.07  0.47  0.532 1.3
Years    0.92 -0.06 -0.05  0.85  0.150 1.0
CAtBat   0.99  0.11  0.04  0.99  0.014 1.0
CHits    0.97  0.13  0.04  0.96  0.037 1.0
CHmRun   0.81  0.28 -0.23  0.80  0.203 1.4
CRuns    0.97  0.17  0.00  0.97  0.026 1.1
CRBI     0.95  0.20 -0.10  0.96  0.037 1.1
CWalks   0.92  0.11 -0.06  0.86  0.138 1.0
PutOuts  0.01  0.34 -0.02  0.12  0.881 1.0
Assists -0.03  0.05  0.94  0.90  0.104 1.0
Errors  -0.11  0.14  0.67  0.48  0.520 1.1

                        MR1  MR2  MR3
SS loadings            6.27 4.55 1.71
Proportion Var         0.39 0.28 0.11
Cumulative Var         0.39 0.68 0.78
Proportion Explained   0.50 0.36 0.14
Cumulative Proportion  0.50 0.86 1.00
```

The first important thing to notice is the factor loadings under the heading "MR1", "MR2", etc. This information indicates how well the variable loads to the factor. The "h2" and "u2" are the communality and uniqueness of each variable. The communality explains how much variance all of the factors explain of the variable. The uniqueness is what is left unexplained. These two categories should sum to one. Low communalities means the factors are not explaining the variable. The last column "com" is a measure of complexity.

As a simple example, the variable `AtBats` has a high loading on MR2 at 0.90 and an overall communality of 0.93, which are both indicators that this variable is well explained. However, the variable `PutOuts` has mostly low loadings on the three factors and has a communality of 0.12 indicating that this variable is not well explained by our factors.

At the bottom of our printout explanations of the sum of squares and variance explained. The model explains almost 80% of the variance which is fairly reasonable. Below is the output for the adequacy test.

CHAPTER 2. EXPLORATORY FACTOR ANALYSIS

```
Mean item complexity =  1.1
Test of the hypothesis that 3 factors are sufficient.

The degrees of freedom for the null model are  120  and the objective function was  30.18 with Chi Square of  7720.79
The degrees of freedom for the model are 75  and the objective function was  7.64

The root mean square of the residuals (RMSR) is  0.04
The df corrected root mean square of the residuals is  0.04

The harmonic number of observations is  263 with the empirical chi square  78.47  with prob <  0.37
The total number of observations was  263  with Likelihood Chi Square =  1939.23  with prob <  0

Tucker Lewis Index of factoring reliability =  0.604
RMSEA index =  0.313  and the 90 % confidence intervals are  0.296 0.32
BIC =  1521.32
Fit based upon off diagonal values = 0.99
```

The main values to interpret are towards the bottom. The Tucker Lewis Index is a measure of the fit this should be near 0.90, which our model fails mightily. The RMSEA should be near 0.08 or lower which is another failure. The Bayesian Information Criterion (BIC) is useful for comparing competing models. Since we only have one model this is not valuable for us in this example. Just as a note, it is common to remove variables in order to improve the results of these indices, remove outliers, and or collect more data.

Now that we have our output we will move to the interpretation of the results. Remember that this more of an art than a science and your persuasive ability is as important as your analytical ability when trying to do this.

Step 5: Interpretation of the Results

For interpretation when need to look at a modified version of the output.This will involve using the print function so we can set the cut off for the loadings we want revealed. By removing low loadings it helps to identify patterns faster. Below is the code and output

```
> print(faResults$loadings,cutoff=0.3)

Loadings:
        MR1    MR2    MR3
AtBat          0.900  0.341
Hits           0.883  0.306
HmRun          0.743
Runs           0.920
RBI            0.901
Walks          0.640
Years   0.918
CAtBat  0.986
CHits   0.972
CHmRun  0.815
CRuns   0.973
CRBI    0.955
CWalks  0.920
PutOuts        0.344
Assists               0.944
Errors                0.669

                MR1    MR2    MR3
SS loadings     6.272  4.545  1.712
Proportion Var  0.392  0.284  0.107
Cumulative Var  0.392  0.676  0.783
```

For me the factors are clear. Factor one is focused on career statistics such as years played cumulative at bats, hits, etc. Factor 2 is focused on the current season offensive statistics. Lastly, factor 3 is focused on the current seasons defensive statistics. The variable `PutOuts` does not seem to load will anywhere and could be removed.

For the final optional step, we will use the factor scores to predict the salaries of the players.

Step 6: Predict with Factor Scores OPTIONAL

To complete this step for predicting the salaries of players we need to do the following

A. Create a dataset with the factor scores.

B. Add the dependent variable to the new dataset

C. Create our train and test sets

D. Train the model

E. Test the model

F. Evaluate the model

To create the dataset with the factor scores and to add the dependent variable to it you can use the code below.

```
fareg<-as.data.frame(faResults$scores)
fareg$Salary<-Hitters$Salary
```

We will now create our train and test set. The code below is the same as in the previous chapter.

```
set.seed(234)
ind<-sample(2,nrow(fareg),replace=T,prob = c(0.7,0.3))
train<-fareg[ind==1,]
test<-fareg[ind==2,]
```

Below is the code for the trained regression model.

```
> model <- lm(Salary~MR1+MR2+MR3,data=train)
> summary(model)

Call:
lm(formula = Salary ~ MR1 + MR2 + MR3, data = train)

Residuals:
    Min      1Q  Median      3Q     Max
-886.17 -177.60   -9.21  102.08 1444.79

Coefficients:
            Estimate Std. Error t value Pr(>|t|)
(Intercept)   516.98      23.37  22.119  < 2e-16 ***
MR1           245.11      23.68  10.351  < 2e-16 ***
MR2           194.86      23.77   8.197 3.44e-14 ***
MR3            25.31      22.67   1.117    0.266
---
Signif. codes:  0 '***' 0.001 '**' 0.01 '*' 0.05 '.' 0.1 ' ' 1

Residual standard error: 326 on 192 degrees of freedom
Multiple R-squared:  0.4996,    Adjusted R-squared:  0.4918
F-statistic:  63.9 on 3 and 192 DF,  p-value: < 2.2e-16
```

The results seem adequate. The last factor is not significant but this is not a major concern. We now need to test our model with the test data. Below is the code.

```
testModel<-predict(model, newdata = test)
```

We will evaluate the model using two metrics. One is the root mean square error and the second is to calculate the correlation between the actual salary values and the predicted ones.

The root mean square error is only useful for model comparison. Therefore, we will compare the train and test results. The calculation of this requires us to subtract the predicted values from the actual ones. Square these values then sum them and finally to square root the final output. The final values should be similar. Below is the code

CHAPTER 2. EXPLORATORY FACTOR ANALYSIS

```
> sqrt(sum((model$fitted.values-train$Salary)^2))
[1] 4516.837
> sqrt(sum((testModel-test$Salary)^2))
[1] 3494.899
```

These are similar. The test model has a lower value because it also has a smaller sample size. Fewer examples means less error.

For our final move we will calculate the correlation between the predict and actual salaries in the test data.

```
> cor(testModel,test$Salary)
[1] 0.365061
> plot(testModel,test$Salary)
```

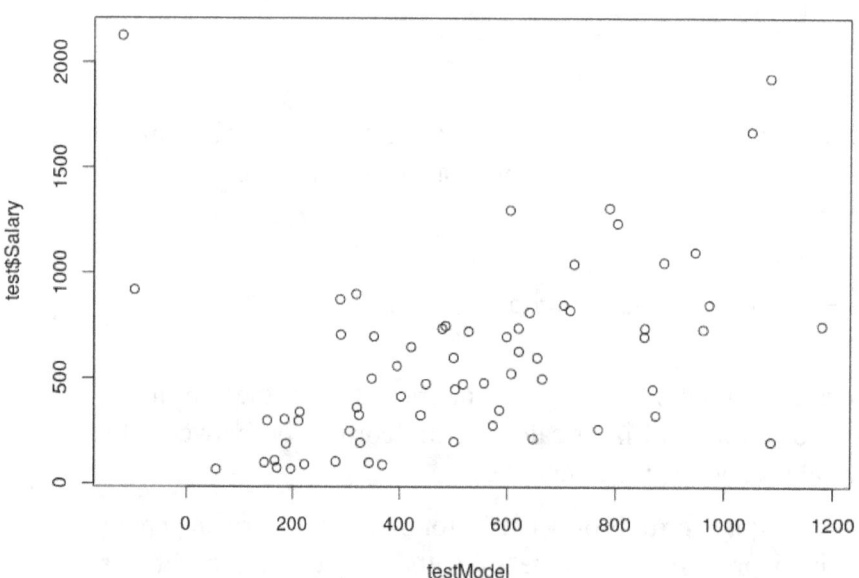

The correlation is low but if you look at the plot there are two extreme values off to the left. If we remove these it would improve the correlation greatly. However, this is test data and we are not suppose to modified the data in the test model.

Conclusion

This chapter gave an explanation and demonstration of the factor analysis. Remember the steps and the subjectivity of the analysis is critical to extracting value from the computation.

Chapter 3

Hierarchical Clustering

Chapter Objectives

- To explain the characteristics of hierarchical clustering.

- To explain the steps involved in conducting hierarchical clustering.

Explaining Hierarchical Clustering

Clustering in general is used to identify patterns in the data. Unlike PCA which is focused on variables or features clustering, hierarchical clustering (HC) is focused on how the examples clump together.

Hierarchical clustering breaks a dataset down into clusters one of two ways and these two approaches are called agglomerative or divisive. Agglomerative has the data begin with each example being an individual cluster that is eventually combined into one giant cluster. Divisive is the opposite. Divisive involves starting with one giant cluster that is broken down into individual examples.

Agglomerative clustering can take a long time to compute with large datasets. Therefore, we will do our analysis with divisive clustering. The steps for our analysis are as follows

1. Data preparation
2. Determine number of clusters
3. Run analysis
4. Interpret the results

Step 1: Data Preparation

For this analysis we will use MedExp data from the Ecdat package. We are trying to identify distinct subgroups in the sample. Below is some initial code with an output of the available variables.

```
> library(Ecdat)
> data("MedExp")
> str(MedExp)
'data.frame':    5574 obs. of  15 variables:
 $ med     : num  62.1 0 27.8 290.6 0 ...
 $ lc      : num  0 0 0 0 0 0 0 0 0 ...
 $ idp     : Factor w/ 2 levels "no","yes": 2 2 2 2
 $ lpi     : num  6.91 6.91 6.91 6.91 6.11 ...
 $ fmde    : num  0 0 0 0 0 0 0 0 0 ...
 $ physlim : Factor w/ 2 levels "no","yes": 1 1 1 1
 $ ndisease: num  13.7 13.7 13.7 13.7 13.7 ...
 $ health  : Factor w/ 4 levels "excellent","good",.
 $ linc    : num  9.53 9.53 9.53 9.53 8.54 ...
 $ lfam    : num  1.39 1.39 1.39 1.39 1.1 ...
 $ educdec : num  12 12 12 12 12 12 12 12 9 9 ...
 $ age     : num  43.9 17.6 15.5 44.1 14.5 ...
 $ sex     : Factor w/ 2 levels "male","female": 1 1
 $ child   : Factor w/ 2 levels "no","yes": 1 2 2 1
 $ black   : Factor w/ 2 levels "yes","no": 2 2 2 2
```

A quick glance show several problems we need to address.

- Reduce the sample size

CHAPTER 3. HIERARCHICAL CLUSTERING

- Remove the categorical varibles

- Scale the data

There is nothing wrong with the size of the dataset. The only problems is that during the writing of this book some of the operations took a long time because of the size of the dataset. To demonstrate this quickly I elected to reduce the size of the sample. Instead of all 5000 we need to use about 1000 speed things up. Therefore, we will randomly pull about 20% of the data for analysis and leave the rest behind. To do this we will slightly modify the code we have been using to create train and test sets. Our new dataset will be called `MedExpSmall`. The code is as follows.

```
set.seed(234)
ind<-sample(2,nrow(MedExp),replace=T,prob = c(0.8,0.2))
toss<-MedExp[ind==1,]
MedExpSmall<-MedExp[ind==2,]
```

We also need to remove the categorical variables because these cannot be analyzed when performing hierarchical clustering. Below is the code to remove the categorical variables.

```
MedExpSmall$sex<-NULL
MedExpSmall$idp<-NULL
MedExpSmall$child<-NULL
MedExpSmall$black<-NULL
MedExpSmall$physlim<-NULL
MedExpSmall$health<-NULL
```

Lastly, hierarchical clustering can be sensitive to unscaled data. It depends but to be safe it is always better to scale the data. When scaling the data we need to store in a data frame object. Below is the code.

```
MedExpSmallScaled<-as.data.frame(scale(MedExpSmall))
```

With the completion of this preparation we are ready to move to the next step which is determining the number of clusters.

Step 2: Determine the Number of Clusters

When conducting a hierarchical clustering you do not have to determine the number of clusters before the analysis. The purpose of determining the number of clusters now is so that we know how to divide our results when we are ready. To decide how many clusters to calculate requires several sub-steps. We have to do the following...

- Calculate the euclidean distance between examples

- Calculate clusters using a linkage method

- Assess the output to determine the number of clusters

Euclidean distance is a way to measure the distance between data points in some sort of multi-dimensional space. The details are a little to complex to explain. The linkage method is the distance between sets of examples as a function of the pairwise distances between examples. If this does not make sense do not worry. For our purposes, we will use Ward's linkage method. However, note that there are other choices available and this will affect the number of clusters recommended.

Once all this is done we will get several outputs. In order to achieve this we will need the `NbClust` function from the `NbClust` package. The arguments inside this function will contain all the information mentioned above. One thing to mention is that you have to set a minimum and maximum number of clusters you want. In addition, you have to tell the function how many different indices you want back. What the function does is it uses over 20 indices to determine the number of clusters and uses simply majority vote to tell you how many to consider. Below is the code and output.

CHAPTER 3. HIERARCHICAL CLUSTERING

```
*******************************************************************
> numComplete<-NbClust(MedExpSmallScaled,distance = 'euclidean',min.nc = 2,
+                     max.nc = 8,method = 'ward.D2',index = c('all'))
*** : The Hubert index is a graphical method of determining the number of clusters.
               In the plot of Hubert index, we seek a significant knee that corresponds to a
               significant increase of the value of the measure i.e the significant peak in Hubert
               index second differences plot.

*** : The D index is a graphical method of determining the number of clusters.
               In the plot of D index, we seek a significant knee (the significant peak in Dindex
               second differences plot) that corresponds to a significant increase of the value of
               the measure.

*******************************************************************
* Among all indices:
* 3 proposed 2 as the best number of clusters
* 5 proposed 3 as the best number of clusters
* 7 proposed 4 as the best number of clusters
* 5 proposed 5 as the best number of clusters
* 3 proposed 8 as the best number of clusters

                    ***** Conclusion *****

* According to the majority rule, the best number of clusters is  4


*******************************************************************
```

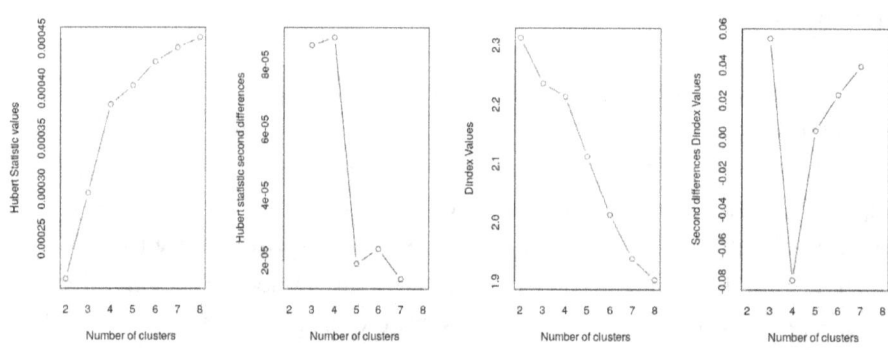

The information at the top is related to the plots at the bottom. They provide a visual of the final recommendation. In the plots you are looking for either a "elbow" or a sudden drop in value. In all four plots you can see the "elbow" or sudden drop in value at around 4 clusters. Above plots is a summary of all the indices that the function calculate. You can see the details of each indices from the code and output below.

```
> numComplete$Best.nc
                    KL      CH Hartigan     CCC    Scott      Marriot
Number_clusters 2.0000   2.000   5.0000  8.0000   3.000 4.000000e+00
Value_Index     8.9708 318.825  24.4491 14.7373 1193.485 1.065246e+25
                 TrCovW  TraceW Friedman   Rubin  Cindex    DB Silhouette
Number_clusters     5.0  5.0000   8.0000  5.0000   3.000 4.0000     4.0000
Value_Index    174684.9 157.1621   4.3313 -0.0239   0.124 1.0329     0.2952
                  Duda PseudoT2   Beale Ratkowsky    Ball PtBiserial Frey
Number_clusters 3.0000   3.0000  2.0000    5.0000   3.000      8.000    1
Value_Index     0.9759  10.4146  1.1571    0.2618 1545.033     0.472   NA
                McClain    Dunn Hubert SDindex Dindex   SDbw
Number_clusters  4.0000  4.0000      0  4.0000      0 4.0000
Value_Index      0.6438  0.1522      0  1.9613      0 0.6311
```

You can clearly see that there are over 20 different indices that all provide various suggestions on the number of clusters to consider. For example, the KL index recommends 2 while the Scott index recommends 3. Various indices use different ways to determine the number of clusters. The tallying of the results is similar to an ensemble method. Through aggregating you can a better sense of what is best.

The conclusion reach was to use 4 clusters. We can now move to the next step and run the actual analysis with 4 clusters.

Step 3: Cluster Analysis

To finally get our clusters we need to calculate the euclidean distance again and store the results. This is done with the `dist` function. Then we can use the `hclust` function to conduct the cluster analysis. There won't be any output yet until we start to make plots. For now, the code below is just to create the clusters.

```
distance<-dist(MedExpSmallScaled,method = 'euclidean')
hiclust<-hclust(distance,method = 'ward.D2')
```

With this we can visualize our clusters using the code below.

```
plot(hiclust,labels=F)
```

CHAPTER 3. HIERARCHICAL CLUSTERING

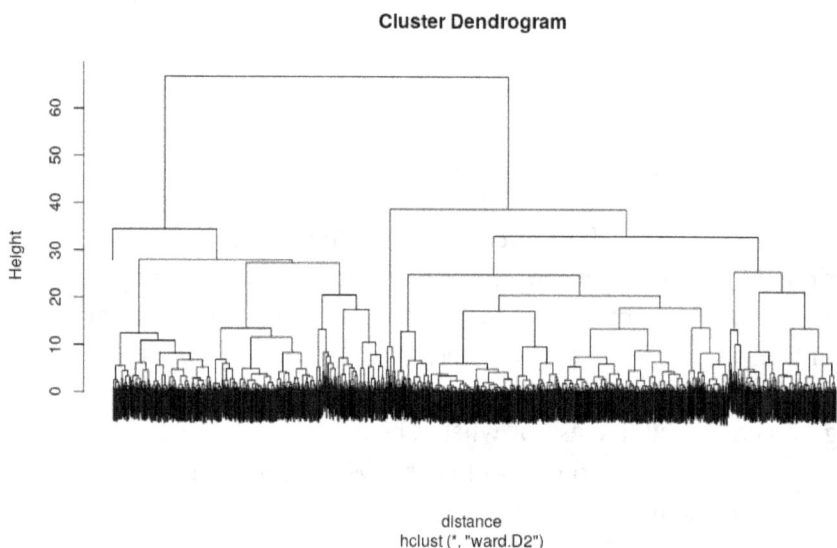

What you see above is know as a dendrogram and is the visual output of a hierarchical cluster analysis. Although it looks like a blob of information there are four clusters in there. They are hard to see. We can try to improve things using the rect.hclust function to highlight the four clusters. The problem is the black and white printing of the book will make it difficult to appreciate the results. Below is the code.

```
rect.hclust(hiclust,k=4,border='gray')
```

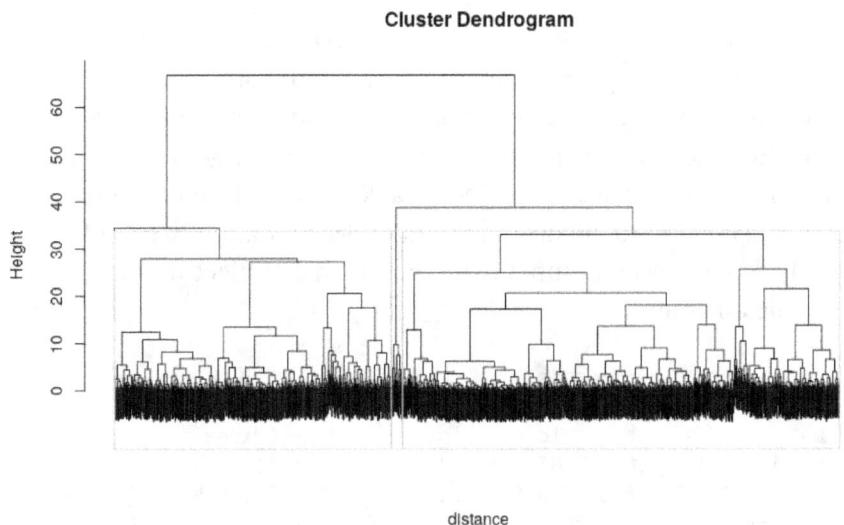

Even with the borders it is hard to see the four clusters. To understand better we can do several forms of descriptive statistics. This will be done in step 4 below.

Step 4: Interpret the Results

First, we will determine how many examples are in each cluster. We do this with the `cutree` function by telling the function how many clusters we want. The code is below.

```
clusters<-cutree(hiclust,k=4)
```

We now take our new `cluster` object and use the `table` function to determine the number of examples in each cluster.

```
> table(clusters)
clusters
  1   2   3   4
422 672  17   1
```

CHAPTER 3. HIERARCHICAL CLUSTERING

Here is our problem. Two of the clusters are extremely small compare to the others. This is why they were hard to see in the dendrogram.

It is also possible to determine descriptive statistics for each cluster using basic functions such as `aggregate` function. We want our unscaled data for descriptive stats because the scaled data is uninterpretable. Therefore, we will add a column to the original `MedExpSmall` dataset we made awhile ago and add the cluster number from the `cluster` object we made. Below is the code and output.

```
> MedExpSmall$clusters<-clusters
> aggregate(MedExpSmall,list(clusters),mean)
  Group.1         med         lc       lpi       fmde ndisease     linc
1       1   115.73410 0.06342805  3.075881 0.06037816 11.72627 8.8158561
2       2   164.73225 4.03872338  5.795219 6.73173524 10.65390 8.7664809
3       3    38.20601 2.72288659  2.644469 4.00160641 10.66726 0.2446402
4       4 17465.98000 0.00000000  5.219274 0.00000000 17.20000 9.6000300
      lfam    educdec      age clusters
1 1.238256 11.821855 24.50195        1
2 1.251135 12.169395 25.09252        2
3 1.123403  9.470588 21.33015        3
4 1.098612 15.000000 54.91924        4
```

This provides us with details on each cluster. For example, people in cluster 3 have the lowest education (9.47) and are also the youngest (21.33). Cluster 4 appears to be an outlier who has serious health problems (med= 17465) and probably should be removed.

Conclusion

Hierarchical clustering provides you with a way to find patterns in the spread of the data. Through doing this you are able to describe various sections of the dataset rather than the entire dataset.

Chapter 4

K-Means Clustering

Chapter Objectives

- To explain the characteristics of k-means clustering.

- To explain the steps involved in conducting k-means clustering.

Explaining K-Means Clustering

K-means clustering is a form of clustering that does not utilize hierarchical methods. Usually, the performance of k-means is normally superior to hierarchical clustering.

K-means works through initializing k number of centroids in the data space. Every example is assigned to the closest centroid. The centroids are then moved until they are in the middle of the examples assigned to them. This process is repeated until the centroids stop moving.

This method is highly flexible, simplistic, and useful for finding patterns in large datasets. However, you do have to set the number of centroids/clusters in advance in order to allow the algorithm to work. This is where the k in k-means comes from. The steps for a k-means analysis are as follows...

1. Data preparation

2. Determine the number of clusters

3. Run the analysis and interpret

4. Predict OPTIONAL

Step 1: Data Preparation

For this chapter, we will use the PSID dataset from the Ecdat package. Below is the initial code.

```
> library(Ecdat)
> data("PSID")
> str(PSID)
'data.frame':   4856 obs. of  8 variables:
 $ intnum  : int  4 4 4 4 5 6 6 7 7 7 ...
 $ persnum : int  4 6 7 173 2 4 172 4 170 171 ...
 $ age     : int  39 35 33 39 47 44 38 38 39 37 ...
 $ educatn : int  12 12 12 10 9 12 16 9 12 11 ...
 $ earnings: int  77250 12000 8000 15000 6500 6500 7000 5000
 $ hours   : int  2940 2040 693 1904 1683 2024 1144 2080 2575
 $ kids    : int  2 2 1 2 5 2 3 4 3 5 ...
 $ married : Factor w/ 7 levels "married","never married",..:
```

CHAPTER 4. K-MEANS CLUSTERING

```
> summary(PSID)
     intnum          persnum            age           educatn         earnings
 Min.   :   4    Min.   :  1.00   Min.   :30.00   Min.   : 0.00   Min.   :     0
 1st Qu.:1905    1st Qu.:  2.00   1st Qu.:34.00   1st Qu.:12.00   1st Qu.:    85
 Median :5464    Median :  4.00   Median :38.00   Median :12.00   Median : 11000
 Mean   :4598    Mean   : 59.21   Mean   :38.46   Mean   :16.38   Mean   : 14245
 3rd Qu.:6655    3rd Qu.:170.00   3rd Qu.:43.00   3rd Qu.:14.00   3rd Qu.: 22000
 Max.   :9306    Max.   :205.00   Max.   :50.00   Max.   :99.00   Max.   :240000
                                                  NA's   :1
     hours            kids               married
 Min.   :   0    Min.   : 0.000   married       :3071
 1st Qu.:  32    1st Qu.: 1.000   never married: 681
 Median :1517    Median : 2.000   widowed       :  90
 Mean   :1235    Mean   : 4.481   divorced      : 645
 3rd Qu.:2000    3rd Qu.: 3.000   separated     : 317
 Max.   :5160    Max.   :99.000   NA/DF         :   9
                                  no histories :  43
```

There are several problems with this dataset that need to be stated.

- There is one missing value in the dataset. This needs to be remove as k-means does not like missing data.

- The `innsum` and `persnum` variables have no meaning as this refers to an interview number or person number. They need to be remove

- The `married` variable is categorical and needs to be removed

- The `kids` and `educatn` variables have values as high as 99. This coding did not mean that someone had 99 kids but was use to represent that the person did not want to share how many kids they had. The same applies for education as nobody goes to school for 99 grades! We need to replace these values with the mean from the column.

- We also must scale our data as k-means is sensitive to this.

- Reduce the size of the dataset for computational reasons.

First, we will remove the missing data with the na.omit function. The code is as follows.

```
PSID<-na.omit(PSID)
```

We will now deal with recoding the 98's and 99's in the educatn and kids variables. We will replace these values with the mean of the column. Below is the code.

```
PSID$educatn[PSID$educatn>=90] <- mean(PSID$educatn)
PSID$kids[PSID$kids>=90] <- mean(PSID$kids)
```

All we are telling R in the above code is to replace any value in the column greater than 90 with the mean.

We also need to remove the married variable as it is a categorical variable and the intnum variable because it is meaningless. Below is the code.

```
PSIDclean<-PSID[,-c(1,2,8)]
```

The -c(1,2,8) in brackets means remove the first, second, and eighth variables which are the intnum, pernum, and married variables. The last two things we need to do is reduce the size of the dataset and scale it. Sample size reduction is only for practical purposes as the computation time for determining the number of clusters can drag on for awhile. In addition, we need to scale our data because k-means is sensitive to this. We will have a sample of about 1200 examples after running the code below.

```
set.seed(234)
ind<-sample(2,nrow(PSIDclean),replace=T,prob = c(0.75,0.25))
toss<-PSIDclean[ind==1,]
PSIDsmall<-PSIDclean[ind==2,]
PSIDscaled<-scale(PSIDsmall)
```

Our new dataset is called PSIDscaled. We are now ready to determine

CHAPTER 4. K-MEANS CLUSTERING

the number of clusters.

Step 2: Determine the Number of Clusters

To figure out how many clusters to use we will use the `NbClust` function from the `NbClust` package. The code and output below should look familiar from the previous chapter. The only different is that the max number of clusters (`max.nc`) was changed from 8 to 12. The code and results are as follows.

```
> library(NbClust)
> numComplete<-NbClust(PSIDscaled,distance = 'euclidean',
+                      min.nc = 2,max.nc = 12,method = 'ward.D2',index = c('all'))
*** : The Hubert index is a graphical method of determining the number of clusters.
            In the plot of Hubert index, we seek a significant knee that corresponds to a
            significant increase of the value of the measure i.e the significant peak in Hubert
            index second differences plot.

*** : The D index is a graphical method of determining the number of clusters.
            In the plot of D index, we seek a significant knee (the significant peak in Dindex
            second differences plot) that corresponds to a significant increase of the value of
            the measure.

*******************************************************************
* Among all indices:
* 8 proposed 2 as the best number of clusters
* 4 proposed 3 as the best number of clusters
* 4 proposed 4 as the best number of clusters
* 2 proposed 5 as the best number of clusters
* 1 proposed 6 as the best number of clusters
* 1 proposed 8 as the best number of clusters
* 1 proposed 9 as the best number of clusters
* 1 proposed 11 as the best number of clusters
* 1 proposed 12 as the best number of clusters

                   ***** Conclusion *****

* According to the majority rule, the best number of clusters is  2

*******************************************************************
```

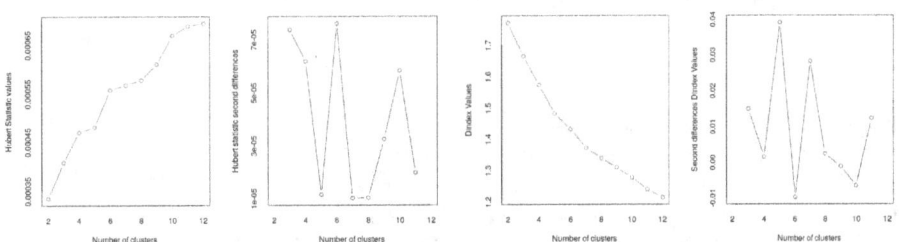

You can see individual index results with the code numComplete$Best.nc. For our purposes we will continue the analyst with 2 clusters.

Step 4: Run Analysis and Interpret

Running the analysis involves the code below. We create an object called clusters with our results and use the kmeans function to complete the analysis

```
clusters<-kmeans(PSIDscaled, 2)
```

You can get an idea of the differences in the two clusters through looking at the cluster center means. Below is the code and output for this.

```
> clusters$centers
         age    educatn    earnings      hours       kids
1 -0.05200781 -0.3907933 -0.7943616 -0.9635865  0.3727023
2  0.03892938  0.2925203  0.5946030  0.7212728 -0.2789786
```

The output indicates strong differences. For example, people in cluster 1 make much less and work much less than people in cluster 2. In addition, people in cluster 1 have way more children.

The only problem with the center results is that they are scaled. This makes it hard to interpret this in the real world. To do that, we need to take the labels our clusters data has and add them to the unscaled data. Our unscaled data was called PSIDsmall. If we add our clusters to this

50

CHAPTER 4. K-MEANS CLUSTERING

dataset we can do another analysis that gives us the unscaled results. Below is the code.

```
> PSIDsmall<-as.data.frame(PSIDsmall)
> PSIDsmall$clusters<-clusters$cluster
> aggregate(PSIDsmall,list(PSIDsmall$clusters),mean)
  Group.1    age  educatn  earnings    hours     kids  clusters
1       1 38.16699 11.37953  2308.961  351.6935 2.767403        1
2       2 38.66765 13.38870 23252.310 1945.2250 1.793960        2
```

Now the data makes more sense. Both groups are about the same age and have the same level of education. However, the difference in earnings, hours worked and number of kids is clearer. The data set is basically divided between the rich and the poor. Our clusters are also about the same size as shown in the code and output below.

```
> table(PSIDsmall$clusters)

  1   2
509 680
```

We can also create a visual using the `clusplot` function from the `cluster` package. This function compresses multiple dimensions of data down to two-dimensions. Now in our case this does not matter as we only have two dimension. However, if you are producing more than two this reduction method allows you to see the data.

To make the plot, we have to use our scaled data again and the cluster results. Below is the code and output.

```
> library(cluster)
> clusplot(PSIDscaled,clusters$cluster,color=TRUE,shade = TRUE,labels=2)
```

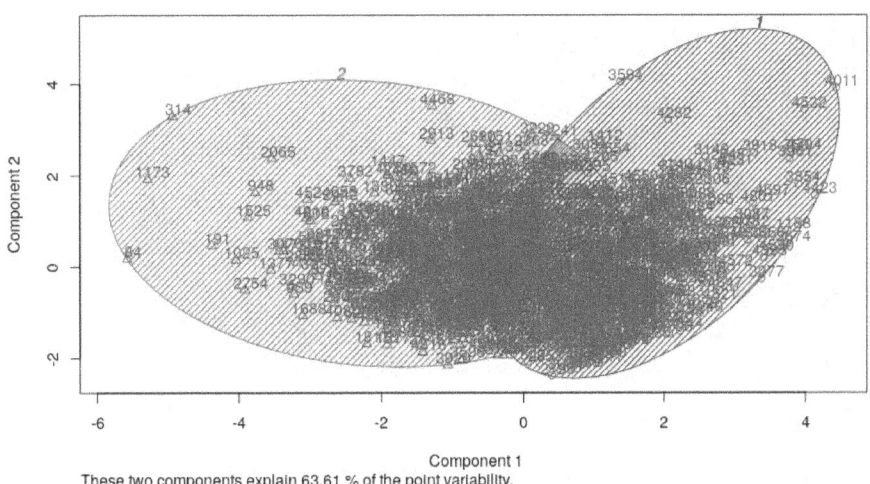

Our interpretation is simply that cluster 1 primarily represents people in poverty while cluster 2 represents people who are middle class or above. You can see that the data is packed close together which is a visual indication that our model is not doing a good job separating the examples.

Predict with Results OPTIONAL

Unlike with hierarchical clustering you can predict with k-means clustering. This can provide insights into the strength of your analysis.

Doing this will require us to do several things

1. Use of the unused data for a test set

2. Predict

3. Assess the accuracy

First we need to make a test set. The test set must be the same size as the train set otherwise there will be problems when we check the accuracy. We will pull the test data from the `toss` dataset we made back at the beginning of the analysis.

```
PSIDtest<-scale(toss[1:1189,])
```

To do the prediction we will use the `cl_predict` function from the `clue` package. After that, we will make our table and calculate our accuracy. Below is the code and output.

```
> library(clue)
> prediction<-cl_predict(clusters,newdata = PSIDtest)
> table(clusters$cluster,prediction)
   prediction
      1   2
  1 201 308
  2 294 386
> mean(clusters$cluster==prediction)
[1] 0.4936922
```

Well, are accuracy is worst than a coin flip. The middle class is there causing problems. You can see this in the clustplot were there is a huge amount of overlap in the two clusters. We made need to try three clusters or consider removing outliers from the existing clusters. Either way, there is still work to be done to improve this model.

Conclusion

This chapter provide an example of the use of k-means clustering. This form of clustering not only allows you to label unlabeled data but can the results can be assessed through supervised learning classification as well. The simplicity and flexibility of this analysis makes it useful.

54

Chapter 5

Mixed Data Clustering

Chapter Objectives

- To explain the characteristics of mixed data clustering.

- To explain the steps involved in conducting mixed data clustering.

Explaining Mixed Data Clustering

Up until this point in book, all of the analyses only utilize continuous variables. However, in the real world, it is often necessary to include categorical data as well.

Mixed Data clustering is a clustering analysis tool that is able to process continuous and categorical data together. The secret to this is the use of the Gower dissimilarity coefficient. The Gower coefficient compares examples pairwise and calculates the dissimilarity between them.

You can use the Gower Coefficient with almost any clustering technique including hierarchical and k-means. Below are the steps.

1. Data preparation

2. Run analysis

3. Interpret results

The steps are fewer in this chapter. The reasons are as follows.

1. We cannot use the `NbClust` to determine the number of clusters because of the presence of categorical variables.

2. We also cannot predict our model accuracy using the `cl_predict` function because of the mixed data.

In this situation you may consider comparing the results of different linkage methods to see if one does a better job than the other. For our purposes we will compare the k-means results with the Partitioning Around the Medoids (PAM). PAM approach minimizes the dissimilarity between examples in cluster. In addition, PAM is more robust to outliers and skewed data when compared to k-means.

Step 1: Date Preparation

We will analyze the `Fair` dataset from the `Ecdat` package. This dataset provides statistics on people who are faithful and unfaithful to their spouse. The data prep is minimal because we do not have to remove categorical variables or create a train/test set. Below is the initial code.

```
library(Ecdata)
data("Fair")
str(Fair)
```

Step 2: Analysis

To do our analysis we need to first calculate the matrix using the Gower coefficient. This requires the use of the `daisy` function from the `cluster` package. The code is below.

```
library(cluster)
gowMat<-daisy(Fair,metric = 'gower')
```

We can now do our analysis. We will do the analysis twice. Once for k-means and once using PAM. We will have three clusters for this analysis. There is no way to include the categorical variables in the analysis to determine how many clusters to create. Often, you have to iterate over the data until you are satisfied.

```
set.seed(234)
clustergow<-pam(gowMat,3)
clusterkm<-kmeans(gowMat,3)
```

We now need to take the label that each example was given from the analysis and add it to our `Fair` dataset. It is best to keep the cluster results separate. Therefore we will duplicate our `Fair` dataset and store the k-means results in that and place the PAM results in the original dataset. Below is the code.

```
Fair1<-Fair #duplicate dataset for kmeans
Fair$clusterg<-clustergow$cluster
Fair1$clusterk<-clusterkm$cluster
```

All we did was create a column in the dataset to store the cluster results for PAM. For k-means we duplicated the dataset and saved it there. We can now move to interpretation.

Step 3: Interpretation

Interpretation is made easy with the use of the `createTable` Function from the `compareGroups` package. The output will provide the descriptive statistics and sample size pf each grow.

First, we need to make the object that the `createTable` function needs. Doing this involves using the `compareGroups` function from the same package. Below is the code.

```
library(compareGroups)
groupg<-compareGroups(clusterg~., Fair) #for gower
groupk<-compareGroups(clusterk~., Fair1) #for kmeans
```

The `~.` symbol means to take all the variables inside the dataset for the aggregation. This is commonly used in regression when you wan to use all the variables in the analysis.

We can now take our two new objects and make the comparison table using the `createTable` function.

```
> createTable(groupg) #forgower

--------Summary descriptives table by 'clusterg'---------

_____
                    1           2           3        p.overall
                  N=235       N=163       N=203
_____
sex:                                                 <0.001
    female    5 (2.13%)  113 (69.3%) 197 (97.0%)
    male    230 (97.9%)   50 (30.7%)   6 (2.96%)
age          35.8 (9.10) 24.7 (4.43) 34.9 (8.76)    <0.001
ym            9.47 (5.10) 2.46 (2.48) 11.3 (4.39)   <0.001
child:                                               <0.001
    no       21 (8.94%) 149 (91.4%)   1 (0.49%)
    yes     214 (91.1%)  14 (8.59%) 202 (99.5%)
religious    3.23 (1.19) 2.68 (1.12) 3.33 (1.09)    <0.001
education    17.5 (2.22) 16.0 (1.85) 14.8 (2.21)    <0.001
occupation   5.25 (1.06) 4.32 (1.45) 2.88 (1.95)    <0.001
rate         3.87 (1.04) 4.36 (0.97) 3.66 (1.18)    <0.001
nbaffairs    1.81 (3.59) 0.76 (2.33) 1.61 (3.53)     0.006
```

```
> createTable(groupk) #for kmeans

--------Summary descriptives table by 'clusterk'---------

_____
                    1           2           3        p.overall
                  N=163       N=216       N=222
_____
sex:                                                 <0.001
    female   99 (60.7%) 216 (100%)    0 (0.00%)
    male     64 (39.3%)   0 (0.00%) 222 (100%)
age          25.4 (5.51) 33.7 (8.44) 36.5 (9.38)    <0.001
ym            2.57 (2.62) 10.5 (4.75) 10.0 (5.03)   <0.001
child:                                               <0.001
    no      162 (99.4%)   0 (0.00%)   9 (4.05%)
    yes       1 (0.61%) 216 (100%)  213 (95.9%)
religious    2.83 (1.15) 3.23 (1.12) 3.22 (1.19)     0.001
education    16.1 (1.94) 15.0 (2.13) 17.3 (2.45)    <0.001
occupation   4.38 (1.43) 3.03 (1.98) 5.19 (1.14)    <0.001
rate         4.31 (0.96) 3.73 (1.18) 3.85 (1.06)    <0.001
nbaffairs    0.82 (2.49) 1.69 (3.56) 1.70 (3.50)     0.015
```

The groupings are remarkable similar. There is a strong separation by gender in both results for 2 of the 3 clusters. The 2nd cluster is much younger (age = 24-25 years) and has almost no children.

It is also possible to create clusterplots using the `clusplot` function from the `cluster` package. Below is the code.

```
clusplot(Fair,clustergow$cluster,
color=TRUE,shade = TRUE,labels=2,main="Gower")
```

CHAPTER 5. MIXED DATA CLUSTERING

```
clusplot(Fair1,clusterkm$cluster,
color=TRUE,shade = TRUE,labels=2,main="Kmeans")
```

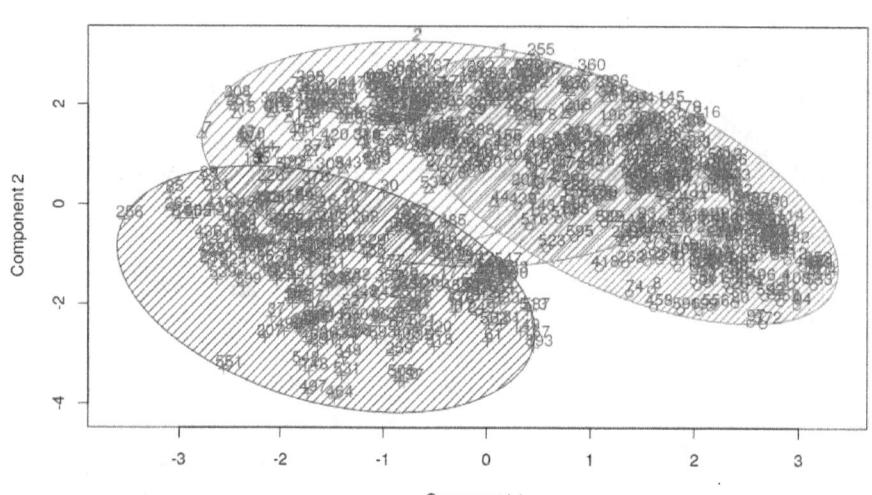

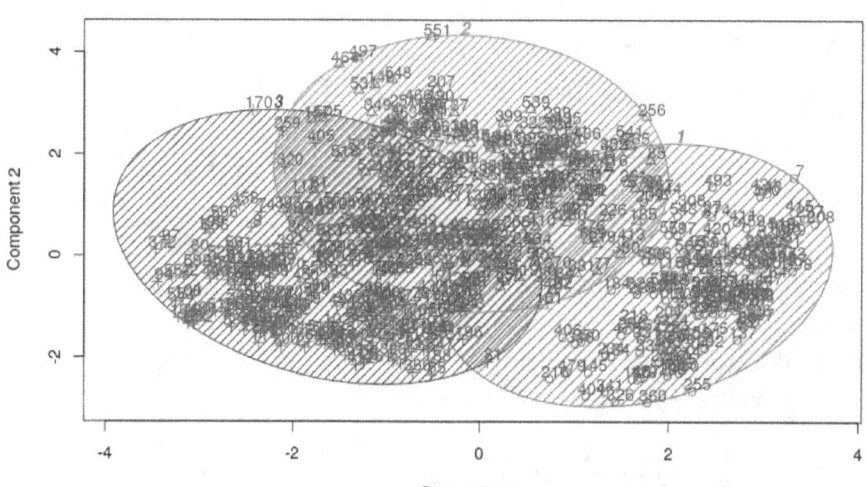

Visualizing the cluster does not looks as impressive but the process is complete and the descriptive statistics revealed differences between the groups.

Conclusion

Mixed data clustering provides a tool to incorporate categorical variables into a clustering analysis. If you have a large number of categorical variables are if you are looking for refined insights into data that is mostly continuous in nature.

Chapter 6

Multi-Dimensional Scaling

Chapter Objectives

- To explain the characteristics of multi-dimensional scaling.

- To explain the steps involved in conducting multi-dimensional scaling.

Explaining Multi-Dimensional Scaling

Multi-dimensional scaling's (MDS) purpose is to plot multivariate data points in two dimensions. This allows us to see what is happening in the data and to identify any potential patterns. This technique was originally used in geography for making maps in space.

MDS is similar to EFA in terms of finding patterns but some forms of MDS can deal with categorical data which EFA cannot. MDS is similar to clustering but there are no clusters and the math is different.

The steps for multi-dimensional scale are as follows

1. Data preparation

2. Analysis and interpretation

In this chapter, there will be two different examples of MDS. The first example will only include metric or continuous data and the second will have mixed or non-metric data.

Metric Data and MDS

Step 1: Data Preparation

We will be using the UScitiesD dataset for our analysis. This dataset gives us the distances that cities are from each other in the form of a matrix.

```
> UScitiesD
              Atlanta Chicago Denver Houston LosAngeles Miami NewYork SanFrancisco Seattle
Chicago           587
Denver           1212     920
Houston           701     940    879
LosAngeles       1936    1745    831    1374
Miami             604    1188   1726     968       2339
NewYork           748     713   1631    1420       2451  1092
SanFrancisco     2139    1858    949    1645        347  2594    2571
Seattle          2182    1737   1021    1891        959  2734    2408          678
Washington.DC     543     597   1494    1220       2300   923     205         2442    2329
```

Step 2: Analysis and Interpretation

The actual scaling is done with the cmdscale function which comes with R. Below is the code for making the scale and the output.

CHAPTER 6. MULTI-DIMENSIONAL SCALING

```
> (mds<-cmdscale(UScitiesD))
                   [,1]        [,2]
Atlanta        -718.7594   142.99427
Chicago        -382.0558  -340.83962
Denver          481.6023   -25.28504
Houston        -161.4663   572.76991
LosAngeles     1203.7380   390.10029
Miami         -1133.5271   581.90731
NewYork       -1072.2357  -519.02423
SanFrancisco   1420.6033   112.58920
Seattle        1341.7225  -579.73928
Washington.DC  -979.6220  -335.47281
```

We can now make our plot. To this we need the `ordiplot` function from the `vegan` package. We set the argument `type` to 't' which stands for 'text'. We also expand the x-axis so that all the text appears on the screen using the `xlim` argument. Below is the code and output.

```
ordiplot(mds,type = 't',xlim=c(-1200,1700))
```

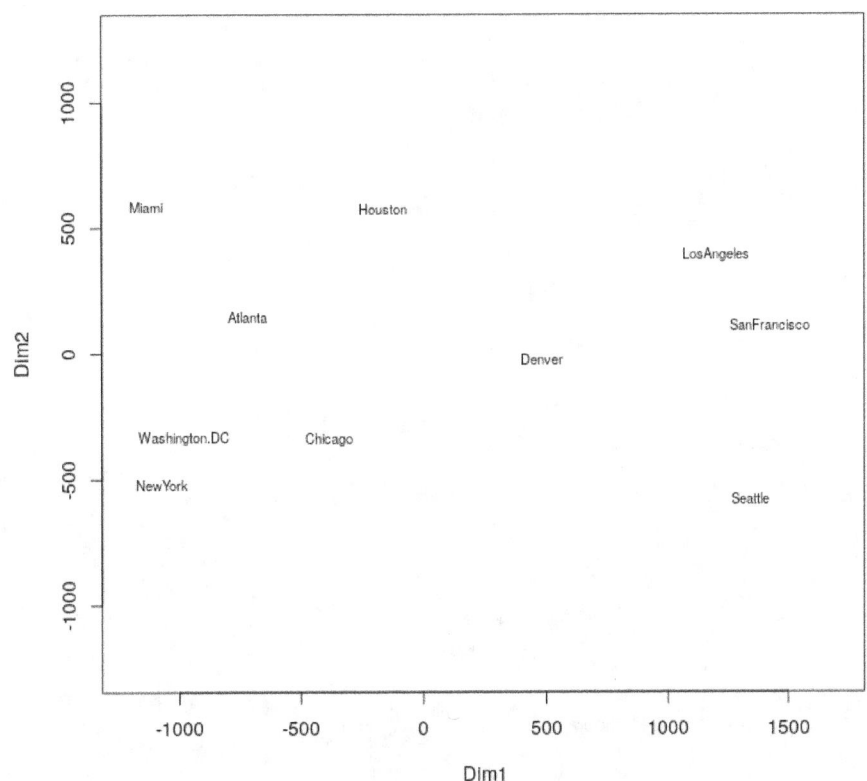

Things look good but something is off. On a map Seattle, San Francisco, and LA would be on your left not on your right. In addition, Seattle is above San Franciso and not below it. For clarification we need to switch the values in our data from positive to negative or vice versa to make the plot clear. To do this you simply multiply the mds dataset by -1. Below is the code and the corrected plot.

```
ordiplot((mds*-1),type = 't',xlim=c(-1700,1200))
```

CHAPTER 6. MULTI-DIMENSIONAL SCALING

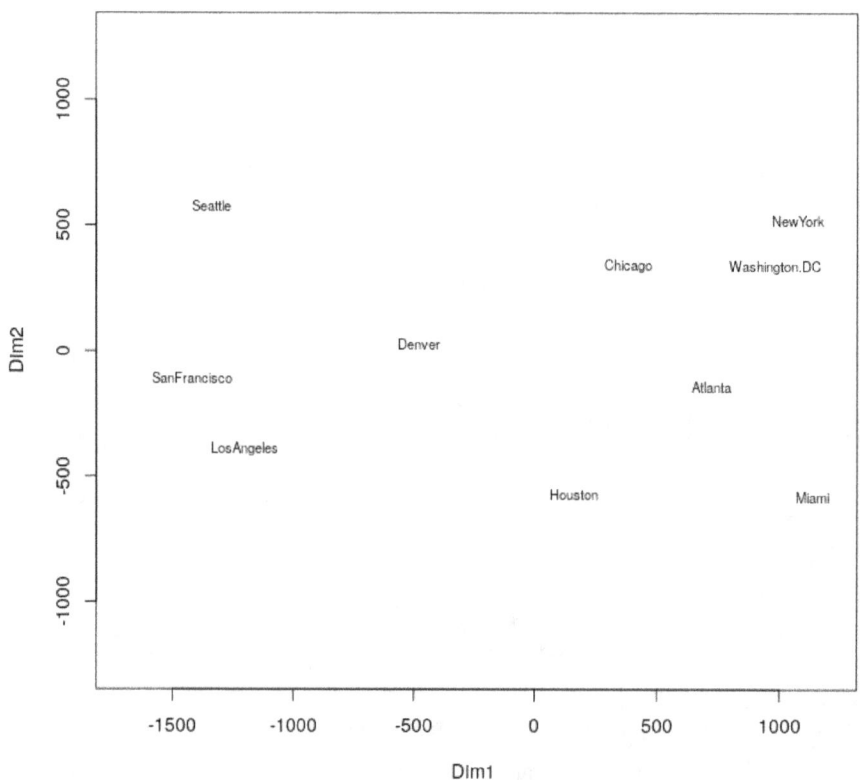

Now this is what we would expect. This mapping of geographic point was one of the original purposes of MDS. We will now turn our attention to working with mixed or non-metric data.

Non-Metric Data and MDS

Step 1: Data Preparation

Our data set for this example is the `College` dataset from the `ISLR` package. Below is the initial code.

```
> library(ISLR)
> data("College")
> str(College)
'data.frame':   777 obs. of  18 variables:
 $ Private    : Factor w/ 2 levels "No","Yes": 2 2
 $ Apps       : num  1660 2186 1428 417 193 ...
 $ Accept     : num  1232 1924 1097 349 146 ...
 $ Enroll     : num  721 512 336 137 55 158 103 489
 $ Top10perc  : num  23 16 22 60 16 38 17 37 30 21
 $ Top25perc  : num  52 29 50 89 44 62 45 68 63 44
 $ F.Undergrad: num  2885 2683 1036 510 249 ...
 $ P.Undergrad: num  537 1227 99 63 869 ...
 $ Outstate   : num  7440 12280 11250 12960 7560 ..
 $ Room.Board : num  3300 6450 3750 5450 4120 ...
 $ Books      : num  450 750 400 450 800 500 500 45
 $ Personal   : num  2200 1500 1165 875 1500 ...
 $ PhD        : num  70 29 53 92 76 67 90 89 79 40
 $ Terminal   : num  78 30 66 97 72 73 93 100 84 41
 $ S.F.Ratio  : num  18.1 12.2 12.9 7.7 11.9 9.4 11
 $ perc.alumni: num  12 16 30 37 2 11 26 37 23 15 .
 $ Expend     : num  7041 10527 8735 19016 10922 ..
 $ Grad.Rate  : num  60 56 54 59 15 55 63 73 80 52
```

We have a few problems to address.

- We have to convert the categorical variable `Private` to a dummy variable.

- The sample size is too large. If we plot all 777 observations our plot will be on giant black blob. MDS works best with small samples so

CHAPTER 6. MULTI-DIMENSIONAL SCALING

you can see individual examples.

To address our first problem we will use the `ifelse` function to recode our `Private` variable.

```
> College$Private<-ifelse(College$Private=="Yes",1,0)
```

To reduce the sample size, we will randomly select 10 colleges from the dataset and save them in the object `CollegeSmall` and we will also print a list of the 10 colleges as well. Below is the code.

```
> set.seed(234)
> CollegeSmall<-College[sample(nrow(College), 10),]
> rownames(CollegeSmall)
 [1] "Taylor University"              "University of California at Irvine"
 [3] "American International College" "Union College KY"
 [5] "Bethel College KS"              "Saint Louis University"
 [7] "Viterbo College"                "St. Norbert College"
 [9] "Virginia Tech"                  "George Mason University"
```

We can now move to the next step.

Step 2: Analysis and Interpretation

To make our matrix we need to use the `vegdist` function from the vegan package. This is a tool commonly used in ecology but can also calculate the distances we need for our purposes. Once this is done we use the `isoMDS` function from the `MASS` package to calculate the coordinates for the MDS plot. Below is the code.

```
> collegedist<-vegdist(CollegeSmall) #calculates distance
> mdsReults<-isoMDS(collegedist) #actual analysis
```

Now we will use the `ordiplot` function again to plot the results.

```
ordiplot(mdsReults$points,type = 't',xlim=c(-.5,.5))
```

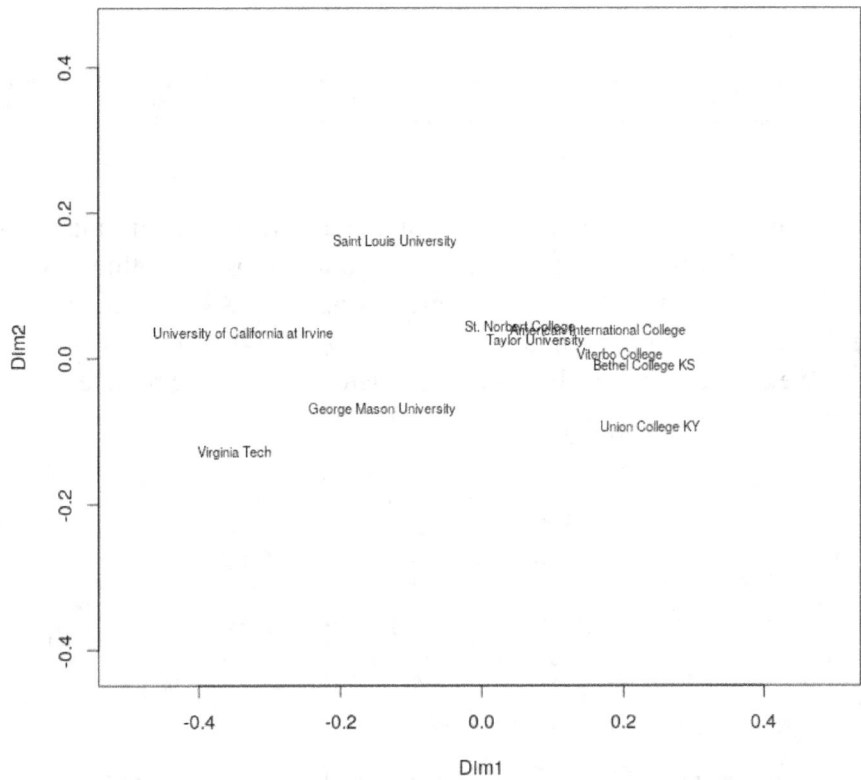

Even with only ten points there is overlap. Imagine with almost 800.

For interpretation, this is more art than science. To try and figure how the examples were plotted I look at extreme values and try to figure out what the differences are. For example, Virginia Tech and Union College KY are as far as possible on the x-axis. If I calculate the difference in the descriptive statistics between these two it gives me an idea of what the x-axis is measuring.

CHAPTER 6. MULTI-DIMENSIONAL SCALING

```
> CollegeSmall[c(9,4),]
                 Private  Apps Accept Enroll Top10perc Top25perc F.Undergrad P.Undergrad Outstate Room.Board Books
Virginia Tech          0 15712  11719   4277        29        53       18511         604    10260       3176   740
Union College KY       1   484    384    177         9        45         634          78     7800       2950   500
                 Personal PhD Terminal S.F.Ratio perc.alumni Expend Grad.Rate
Virginia Tech        2200  85       89      13.8         20   8944        73
Union College KY      600  60       88      14.1          9   6864        64
> CollegeSmall[9,]-CollegeSmall[4,]
              Private  Apps Accept Enroll Top10perc Top25perc F.Undergrad P.Undergrad Outstate Room.Board Books
Virginia Tech      -1 15228  11335   4100        20         8       17877         526     2460        226   240
              Personal PhD Terminal S.F.Ratio perc.alumni Expend Grad.Rate
Virginia Tech     1600  25        1      -0.3         11   2080         9
```

By subtracting the two values from each other you can see that Virginia Tech has much higher metrics than Union College in almost everything. Therefore, it is possible that the x-axis is measuring size. As you move to the left the schools get bigger and as you move to the right the schools get smaller. We can confirm this by sorting the data frame by the number of freshman.

```
> CollegeSmall[order(CollegeSmall$F.Undergrad),]
                                  Private  Apps Accept Enroll Top10perc Top25perc F.Undergrad P.Undergrad
Bethel College KS                       1   202    184    122        19        42         537         101
Union College KY                        1   484    384    177         9        45         634          78
Viterbo College                         1   647    518    271        17        43        1014         387
American International College          1  1420   1093    220         9        22        1018         287
Taylor University                       1  1769   1092    437        41        80        1757          81
St. Norbert College                     1  1334   1243    568        30        56        1946          95
Saint Louis University                  1  3294   2855    956        44        67        4576        1140
George Mason University                 0  5653   4326   1727        17        29        9528        3822
University of California at Irvine      0 15698  10775   2478        85       100       12677         864
Virginia Tech                           0 15712  11719   4277        29        53       18511         604
```

You can clearly see that Union College is one of the smallest in ermfo Apps, Accept, Enroll, F.Undergrad, and P.Undergrad while Virigina Tech is one of the largest.

We repeat this process for the y-axis. The highest and lowest schools are Saint Louis University and Virginia Tech.

```
> CollegeSmall[c(9,6),]
                       Private  Apps Accept Enroll Top10perc Top25perc F.Undergrad P.Undergrad Outstate Room.Board
Virginia Tech                0 15712  11719   4277        29        53       18511         604    10260       3176
Saint Louis University       1  3294   2855    956        44        67        4576        1140    11690       4730
                       Books Personal PhD Terminal S.F.Ratio perc.alumni Expend Grad.Rate
Virginia Tech            740     2200  85       89      13.8         20   8944        73
Saint Louis University   800     6800  84       94       4.6         19  18367        67
> CollegeSmall[9,]-CollegeSmall[6,]
              Private  Apps Accept Enroll Top10perc Top25perc F.Undergrad P.Undergrad Outstate Room.Board Books
Virginia Tech      -1 12418   8864   3321       -15       -14       13935        -536    -1430      -1554   -60
              Personal PhD Terminal S.F.Ratio perc.alumni Expend Grad.Rate
Virginia Tech    -4600   1       -5       9.2          1  -9423         6
```

The difference here appears to be financial. Saint Louis University cost much more and spends a lot more on their students compared to Virginia Tech. If we sort by expend we get a better insight.

```
CollegeSmall[order(CollegeSmall$Expend),]
```

	Room.Board	Books	Personal	PhD	Terminal	S.F.Ratio	perc.alumni	Expend
George Mason University	4840	580	1050	93	96	19.3	7	6751
Union College KY	2950	500	600	60	88	14.1	9	6864
American International College	4780	450	1400	78	84	14.7	19	7355
Viterbo College	3365	500	2245	51	65	10.7	31	8050
Taylor University	4000	450	1250	60	61	14.2	32	8294
Bethel College KS	3580	500	1400	61	80	8.8	32	8324
St. Norbert College	4450	425	1100	74	78	15.1	36	8595
Virginia Tech	3176	740	2200	85	89	13.8	20	8944
University of California at Irvine	5302	790	1818	96	96	16.1	11	15934
Saint Louis University	4730	800	6800	84	94	4.6	19	18367

Saint Louis University is clearly number one but Virginia Tech is not the last. Saint Louis is just way more expensive then the pack. Another major difference is the variable `Personal`. Perhaps you can see why have a large dataset would be difficult to use when you want to understand the results and just map them. This is just an example of how to understand MDS.

Conclusion

Multi-dimensional scale allows you to map data points with many variables into a two-dimensional space. This allows you to see how the data points relate to each other and to develop hypotheses for why they are similar or different.

Chapter 7

Market Basket Analysis

Chapter Objectives

- To explain the characteristics of market basket analysis.

- To explain the steps involved in conducting market basket analysis.

Explaining Market Basket Analysis

Market basket analysis is a machine learning approach that attempts to find relationships among a group of items in a data set. For example, a supposedly famous use of this method was when one retailer found an association between beer and diapers. Upon closer examination, the retailers found that when men came to purchase diapers for their babies they would often buy beer at the same time. With this knowledge, the retailer placed beer and diapers next to each other in the store and this helped to further increase sales.

In addition, many of the recommendation systems we experience when shopping online use market basket analysis results to suggest additional products. As such, market basket analysis is an intimate part of our lives without us even knowing.

At the heart of market basket analysis are association rules. Association rules explain patterns of relationship among items. Below is an example.

{rice, seaweed} -> {soy sauce}

Everything inside curly braces { } is an itemset. An itemset is some form of data that occurs frequently in the dataset based on some criteria. Rice and seaweed are our itemset on the left and soy sauce is our itemset on the right. The arrow -> indicates what comes first as we read from left to right. If we stated this association rule in plain English it would say "if a person buys rice and seaweed then they will buy soy sauce". The practical use of this rule is to place rice, seaweed and soy sauce near each other in the store in order to encourage the purchase of them when people come to shop.

Market basket analysis uses an apriori algorithm. This algorithm is useful for unsupervised learning that does not require any training and thus no predictions. The apriori algorithm is especially useful with large datasets but it employs simple procedures to find useful relationships among the items. The shortcut that this algorithm uses is the âĂIJapriori propertyâĂİ which states that all suggests of a frequent itemset must also be frequent. What this means in simple English is that the items in an itemset need to be common in the overall dataset. This simple rule saves a tremendous amount of computational time.

The steps for completing a market basket analysis are as follows...

1. Data exploration

2. Analysis

Step 1: Data Exploration

We will be using the `Groceries` dataset from the `arules` package. The arules packages is also the package we will use to do the analysis. Be-

CHAPTER 7. MARKET BASKET ANALYSIS

low is some initial code.

```
library(arules)
data("Groceries")
```

The data is already clean and ready to go. Therefore, we will explore it briefly before conducting the analysis. The exploration can help you to determine what you want to know if you do not know beforehand. Below is a summary of the dataset.

```
> summary(Groceries)
transactions as itemMatrix in sparse format with
 9835 rows (elements/itemsets/transactions) and
 169 columns (items) and a density of 0.02609146

most frequent items:
      whole milk other vegetables       rolls/buns            soda          yogurt         (Other)
            2513            1903             1809            1715            1372           34055

element (itemset/transaction) length distribution:
sizes
   1    2    3    4    5    6    7    8    9   10   11   12   13   14   15   16   17   18   19   20
2159 1643 1299 1005  855  645  545  438  350  246  182  117   78   77   55   46   29   14   14    9

   Min. 1st Qu.  Median    Mean 3rd Qu.    Max.
  1.000   2.000   3.000   4.409   6.000  32.000

includes extended item information - examples:
       labels  level2           level1
1 frankfurter sausage meat and sausage
2     sausage sausage meat and sausage
3  liver loaf sausage meat and sausage
```

The output tells us the number of rows in our dataset (9835) columns (169) as well as the density. The density is the percentage of columns that are not empty (2.6%). This may seem small but remember that the number of purchases varies from person to person so this affects how many empty columns there are. This is actually common when creating sparse matrices.

Next, we have the most commonly purchased items. Milk and other vegetables were the two most common followed by other foods. After the most frequent items, we have the size of each transaction. For example, 2159 people purchased one item during a transaction. While 20 people purchased 9 items in a transaction. Lastly, we have some summary statistics about transactions. On average, a person would purchase around 4.4 items per transaction.

In order to create a visual of the most commonly purchased items we will use the `itemFrequencyPlot` function and we will add the argument `topN` to sort the items from most common to least for the 15 most frequent transactions. Below is the code and the output.

```
itemFrequencyPlot(Groceries,topN=15)
```

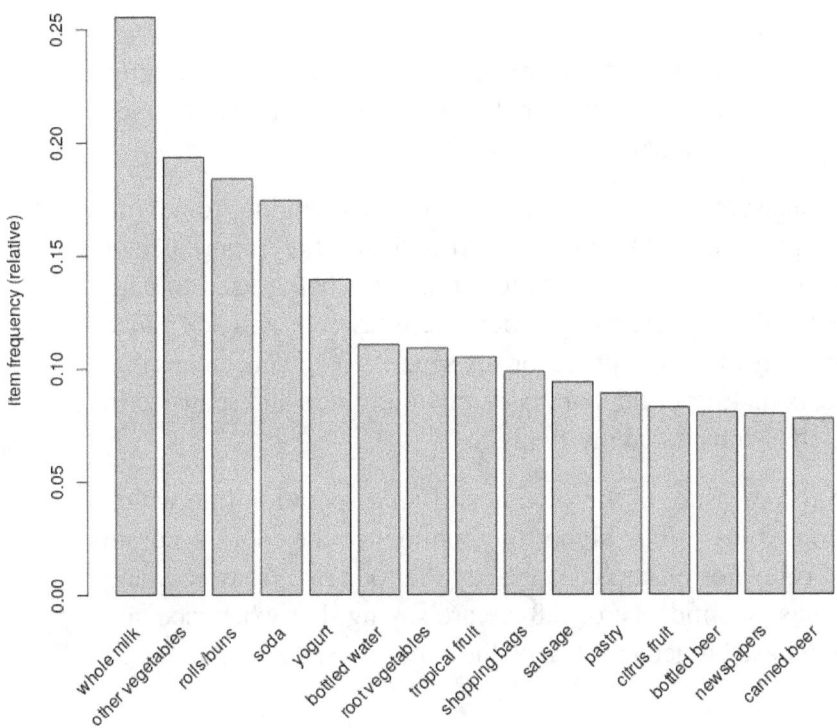

We already knew the top five from the `summary` output but now we know the top 15 items. Whole milk appears in about 25% of the purchases, other vegetables in about 20%, etc. The bar graph also gives you an idea of the proportions of each. For example, after yogurt the frequen-

CHAPTER 7. MARKET BASKET ANALYSIS

cies are all about the same.

This is enough for the exploration aspect of this chapter. We will now develop our association rules in the next step.

Step 2: Analysis

The analysis requires the use of the `apriori` function. Before we use this function we need to go over support and confidence as they are defined when doing a market basket analysis.

Support is a measure of the frequency of an itemset ranging from 0 (no support) to 1 (highest support). High support indicates the importance of the itemset in the data and contributes to the itemset being used to generate association rule(s). Returning to our rice, seaweed, and soy sauce example. We can say that the support for soy sauce is 0.4. This means that soy sauce appears in 40% of the purchases involving rice and seaweed in the dataset which is considered pretty high.

Confidence is a measure of the accuracy of an association rule which is measured from 0 to 1. The higher the confidence the more accurate the association rule. For example, If we say that our rice, seaweed, and soy sauce rule has a confidence of 0.8 we are saying that when rice and seaweed are purchased together, 80% of the time soy sauce is purchased as well.

The default settings for the `apriori` function for support is 0.1 and for confidence 0.8. This means that for a rule to be created the item must appear in 10%of the transactions and be 80% accurate. Below is the code and output.

```
> apriori(Groceries)
Apriori

Parameter specification:
 confidence minval smax arem  aval originalSupport maxtime support minlen maxlen target   ext
        0.8    0.1    1 none FALSE            TRUE       5     0.1      1     10  rules FALSE

Algorithmic control:
 filter tree heap memopt load sort verbose
    0.1 TRUE TRUE  FALSE TRUE    2    TRUE

Absolute minimum support count: 983

set item appearances ...[0 item(s)] done [0.00s].
set transactions ...[169 item(s), 9835 transaction(s)] done [0.00s].
sorting and recoding items ... [8 item(s)] done [0.00s].
creating transaction tree ... done [0.00s].
checking subsets of size 1 2 done [0.00s].
writing ... [0 rule(s)] done [0.00s].
creating S4 object  ... done [0.00s].
set of 0 rules
```

The initial analysis is not promising. If you look at the bottom of the output no rules were produced. This means nothing met our criteria of 0.1 support and 0.8 confidence. In order to produce rules we need to manipulate these parameters.

How to play with these numbers is a matter of experience as there are few strong rules for this matter. Below, I set the support to 0.03, confidence to 0.25, and the minimum number of rule items to 2. By setting the minimum number of items to two this means that at a minimum there can be one item on the left and one item on the right in the rule. Keep in mind that this is minimum so there can be many more items on other side of the rule

Once the rules are made we sort them by lift. Lift is the support divided by the product of the probabilities of the items on the left and right hand side happening as if there was no association between them. A lift greater than 1 suggests indicates how much better a rule is than just make an assumption and is also a way to summarize the strength of a rule. The higher the lift the stronger the association between the itemsets in the transaction.

Below is the code with the output as well.

CHAPTER 7. MARKET BASKET ANALYSIS

```
groceriesrules<-apriori(groceries, parameter = list(support=0.03,
confidence = 0.25, minlen=2))
> groceriesrules
set of 15 rules
> inspect(sort(groceriesrules, by="lift")[1:7])
    lhs                       rhs                support    confidence lift     count
[1] {root vegetables}      => {other vegetables} 0.04738180 0.4347015  2.246605 466
[2] {sausage}              => {rolls/buns}       0.03060498 0.3257576  1.771048 301
[3] {tropical fruit}       => {other vegetables} 0.03589222 0.3420543  1.767790 353
[4] {whipped/sour cream}   => {whole milk}       0.03223183 0.4496454  1.759754 317
[5] {root vegetables}      => {whole milk}       0.04890696 0.4486940  1.756031 481
[6] {yogurt}               => {other vegetables} 0.04341637 0.3112245  1.608457 427
[7] {tropical fruit}       => {whole milk}       0.04229792 0.4031008  1.577595 416
```

The output indicates that 15 rules were created. Of these 15, we have printed out the top 7 as sorted by lift. Below is what the first three rules mean in simply English.

1. People who buy root vegetables also buy other vegetables. This occurs 4.7% of the time in the dataset and the rule is 43% accurate. If root vegetables are purchased people are 2.2 times more likely to buy other vegetables.

2. People who buy sausage also buy rolls/buns. This occurs in 3% of the transactions and the rule is 33% accurate. If sausage is purchased people are 1.8 times more likely to purchase rolls/buns.

3. People who buy tropical fruit also buy other vegetables. This occurs in 3.6% of the transactions and is 34% accurate. If people purchase tropical fruit they are 1.8 times more likely to purchase other vegetables.

It is also possible to make visuals of association rules. To do this you need the `plot` function from the `arulesViz` package. We will make a graph of our fifteen rules in the code below.

```
library(arulesViz)
plot(groceriesrules,method = "graph")
```

The key in the upper right tells you that the size of the bubble indicates the support and the color the lift. Unfortunately, in a black and white text it is impossible to appreciate the lift characteristic in this visual.

You can adjust the amount of support, confidence, and the number of items purchased to create different rules than these. The process of creating rules is highly subjective. The key is to maintain simplicity and to have some sort of justification/explanation for the steps you take to generate rules.

Conclusion

Market basket analysis is a clue for determining how various items are related in a dataset. For business reasons, this approach can be used to determine were to place items in a store. This can have a powerful financial effect on a business.

80

Chapter 8

K-Nearest Neighbor

Chapter Objectives

- To explain the characteristics of k-nearest neighbor.

- To explain the steps involved in conducting k-nearest neighbor.

Explaining K-Nearest Neighbor

K-Nearest neighbor /(KNN/)is an approach that uses labeled data that is near an unlabeled example to determine what the unknown example class or value is. The thought is that similar classes or predictions should be in the same feature space. This is an example of supervised learning because the examples are clearly labeled and this information is used to determine the values of the unlabeled data. As this is supervised, the approach is partially different as we have to have a train and test set to develop our model. This will be explained in detail later

There are two types of KNN one for classification and the other for numeric prediction or regression. This chapter contains one example of each.

KNN classification uses a simple technique to classify unlabeled examples. As mentioned before, the algorithm assigns an unlabeled example the label of the nearest example(s). This based on the assumption that if two examples are next to each other they must be of the same class.

KNN classification uses the features of the data set to create a multidimensional feature space. The number of variables decides the number of dimensions in the space. Therefore, two variables leads to a two-dimensional feature space, three variables leads to a three dimensional feature space, etc. Within this feature space all the examples are placed based on their respective values of the class they belong to.

The label of the unknown examples are determined by who the closet neighbor(s) is or are. This calculation is based on Euclidean distance. The number of neighbors that are used to calculate the distance varies at the discretion of the researcher. The "K" in KNN stands for the number of neighbors that are used by the algorithm. For example, we could use one neighbor or several to determine the label of an unlabeled example. There are advantages and disadvantages in determining how many neighbors to use. In general, the more neighbors used the more complicated the classification becomes.

Nearest neighbor classification is considered a type of lazy learning. What is meant by lazy is that no abstraction of the data happens. This means there is no real explanation or theory provided by the model to infer why there are certain relationships in the data. Nearest neighbor tells you where the relationships are but not why or how. This is partly due to the fact that it is a non-parametric learning method and provides no parameters (summary statistics) about the data.

Below are the steps we will employ for KNN analysis.

1. Data preparation

2. Model training

3. Model testing

KNN for Classification

Step 1: Data Preparation

For our classification KNN, we will use the `Mroz` dataset from the `Ecdat` package. Our goal will be to predict if someone lives in a large city or not which is our two category categorical variable. Below is some initial code with an overview of the types of variables in the dataset.

```
> library(Ecdat)
> data("Mroz")
> str(Mroz)
'data.frame':   753 obs. of  18 variables:
 $ work      : Factor w/ 2 levels "yes","no": 2 2 2
 $ hoursw    : int  1610 1656 1980 456 1568 2032 14
 $ child6    : int  1 0 1 0 1 0 0 0 0 ...
 $ child618  : int  0 2 3 3 2 0 2 0 2 2 ...
 $ agew      : int  32 30 35 34 31 54 37 54 48 39 .
 $ educw     : int  12 12 12 12 14 12 16 12 12 12 .
 $ hearnw    : num  3.35 1.39 4.55 1.1 4.59 ...
 $ wagew     : num  2.65 2.65 4.04 3.25 3.6 4.7 5.9
 $ hoursh    : int  2708 2310 3072 1920 2000 1040 2
 $ ageh      : int  34 30 40 53 32 57 37 53 52 43 .
 $ educh     : int  12 9 12 10 12 11 12 8 4 12 ...
 $ wageh     : num  4.03 8.44 3.58 3.54 10 ...
 $ income    : int  16310 21800 21040 7300 27300 19
 $ educwm    : int  12 7 12 7 12 14 14 3 7 7 ...
 $ educwf    : int  7 7 7 7 14 7 7 3 7 7 ...
 $ unemprate : num  5 11 5 5 9.5 7.5 5 5 5 3 5 ...
 $ city      : Factor w/ 2 levels "no","yes": 1 2 1
 $ experience: int  14 5 15 6 7 33 11 35 24 21 ...
```

There are several things we need to do in to prepare the data.

- Remove the `work` variable because it is categorical and KNN cannot handle this as one of the independent variables.

- Scale the data.

- Create the train and test data.

This is not that hard to do and we can actually handle the first two bullets with a few lines of code. Below is the code and the output.

```
> Mroz$work<-NULL
> mroz.scale<-as.data.frame(scale(Mroz[,-16]))
> mroz.scale$city<-Mroz$city
> str(mroz.scale)
'data.frame':   753 obs. of  17 variables:
 $ hoursw     : num  0.998 1.051 1.422 -0.327 0.95 ...
 $ child6     : num  1.455 -0.454 1.455 -0.454 1.455 ...
 $ child618   : num  -1.03 0.49 1.25 1.25 0.49 ...
 $ agew       : num  -1.305 -1.553 -0.934 -1.058 -1.429
 $ educw      : num  -0.126 -0.126 -0.126 -0.126 0.751 .
 $ hearnw     : num  0.302 -0.304 0.67 -0.394 0.684 ...
 $ wagew      : num  0.331 0.331 0.905 0.579 0.723 ...
 $ hoursh     : num  0.74 0.0717 1.3512 -0.5831 -0.4488
 $ ageh       : num  -1.38 -1.876 -0.635 0.978 -1.628 ..
 $ educh      : num  -0.163 -1.156 -0.163 -0.825 -0.163
 $ wageh      : num  -0.816 0.227 -0.922 -0.931 0.595 ..
 $ income     : num  -0.555 -0.105 -0.167 -1.295 0.346 .
 $ educwm     : num  0.816 -0.668 0.816 -0.668 0.816 ...
 $ educwf     : num  -0.506 -0.506 -0.506 -0.506 1.453 .
 $ unemprate  : num  -1.163 0.763 -1.163 -1.163 0.281 ..
 $ experience : num  0.418 -0.698 0.541 -0.574 -0.45 ...
 $ city       : Factor w/ 2 levels "no","yes": 1 2 1 1 2
```

We will now create our train and test datasets. This is done for the purpose of developing a model with the train dataset and assessing the model's strength with the test dataset.

```
set.seed(234)
ind=sample(2,nrow(mroz.scale),replace=T,prob=c(.7,.3))
```

```
train<-mroz.scale[ind==1,]
test<-mroz.scale[ind==2,]
```

We can now proceed to model development

Step 2: Model Development

To improve our model we are going to use a concept called cross-validation. Cross-validation involves comparing your model results by how it does on new data. We have done this several times already.

However, in this chapter we are going to use k-fold cross validation. K-fold involves folding your training dataset k times. K - 1 of the folds are used to train your model and the last fold is used to evaluate. The results of the evaluation are averaged to create the final estimates for the model.

To set up the cross-validation we will need the `trainControl` function from the `caret` package. For now we need to create an object that we will use in the near future.

```
library(caret)
control<-trainControl(method="cv")
```

Another problem we need to address is we do not know how many neighbors to use in order to label our data in the dataset. Remember this is k nearest neighbor. We must determine how large k needs to be in order to complete our analysis. The only solution is random guessing and let the computer decide what is the appropriate number. Therefore, we will create a grid that goes from 2 - 17. This means that R will create 16 models one for 2,3,4,5 etc until 17. R will then tell us what is the most accurate value for k. Below is the code.

```
grid<-expand.grid(.k=seq(2,25,by=1))
```

We are now ready to run our model. This time we will run the code and explain it afterwards.

```
> knn.train<-train(city~.,train,method="knn",trControl=control,tuneGrid=grid)
> knn.train
k-Nearest Neighbors

561 samples
 16 predictor
  2 classes: 'no', 'yes'

No pre-processing
Resampling: Cross-Validated (10 fold)
Summary of sample sizes: 505, 505, 505, 505, 505, 505, ...
Resampling results across tuning parameters:

   k   Accuracy   Kappa
   2   0.6220865  0.1825579
   3   0.6257832  0.1437029
   4   0.6347431  0.1687600
   5   0.6417607  0.1658215
   6   0.6328321  0.1436718
   7   0.6364662  0.1284997
   8   0.6595551  0.1881280
   9   0.6649123  0.1932490
  10   0.6631579  0.1862760
  11   0.6738095  0.2043984
  12   0.6862782  0.2310447
  13   0.6933897  0.2522674
  14   0.6863095  0.2335871
  15   0.6827068  0.2197405
  16   0.6826754  0.2160756
  17   0.6755326  0.1996389

Accuracy was used to select the optimal model using the largest value.
The final value used for the model was k = 13.
```

Here is what the code means for our `knn.train` object from left to right.

1. We used the `train` function from the `caret` package

2. We created the formula

3. We indicated the dataset which is also called `train`

4. The type of model being trained is indicated with the `method` argument. knn stands for k nearest neighbor.

5. `trControl` is the argument we used to setup the cross-validation. We made the object called `control` for this purpose.

6. The argument `tunerGrid` is where we placed the `grid` object we created earlier.

The output tells you the sample size, predictors, and number of classes. After this, it provides some details about the cross-validation. Then the results for each number of k neighbors is listed with the accuracy and kappa. Kappa is a metric of accuracy that controls for random guessing. The values here are pretty poor. At the bottom of the printout R recommends 13 as our k. This is based solely on the accuracy.

Step 3: Model Testing

Now that we know how many neighbors to use we can now run the model with the k set. To actual do the analysis we need the `knn` function from the `class` package. Below is the code with the output.

```
> library(class)
> knn.test<-knn(train[,-17],test[,-17],train[,17],k=13)
> table(knn.test,test$city)

knn.test  no  yes
     no   22   10
     yes  45  115
> mean(knn.test==test$city)
[1] 0.7135417
> kappa(table(knn.test,test$city))
         Estimate  Std.Err   2.5%   97.5%  P-value
kappa    0.2826   0.06862  0.1481  0.4171  3.81e-05
```

Here is what happened.

1. For the `knn.test` object we had to remove the 17th variable because that is the dependent variable and we wanted to predict this. This is the reason for the -17 in brackets. Notice also that the k is set to 13 based on our prior analysis

2. The `table` function shows us the results and the accuracy is calculated below it. 71% is reasonable.

3. the `kappa` function was used to calculated the kappa. .28 is not that great

There is one final trick we can use and this has to do with determining if weighted or unweighted k neighbors should be use. The example we just did was with unweighted k neighbors. There are times when weighted neighbors can improve accuracy. With weighted KNN the neighbor that is the closest is given a stronger weight or greater emphasis in he classification or regression.

We will look at three different weighing methods. "Rectangular" is unweighted and is the one that we used. The other two are "triangular" and "epanechnikov." How these calculate the weights is beyond the scope of this book. In the code below the argument "distance" can be set to 1 for Euclidean and 2 for absolute distance.

```
> kknn.train<-train.kknn(city~.,train,kmax = 25,distance = 2,kernel = c("rectangular","triangular",
+                                                                         "epanechnikov"))
> plot(kknn.train)
```

CHAPTER 8. K-NEAREST NEIGHBOR

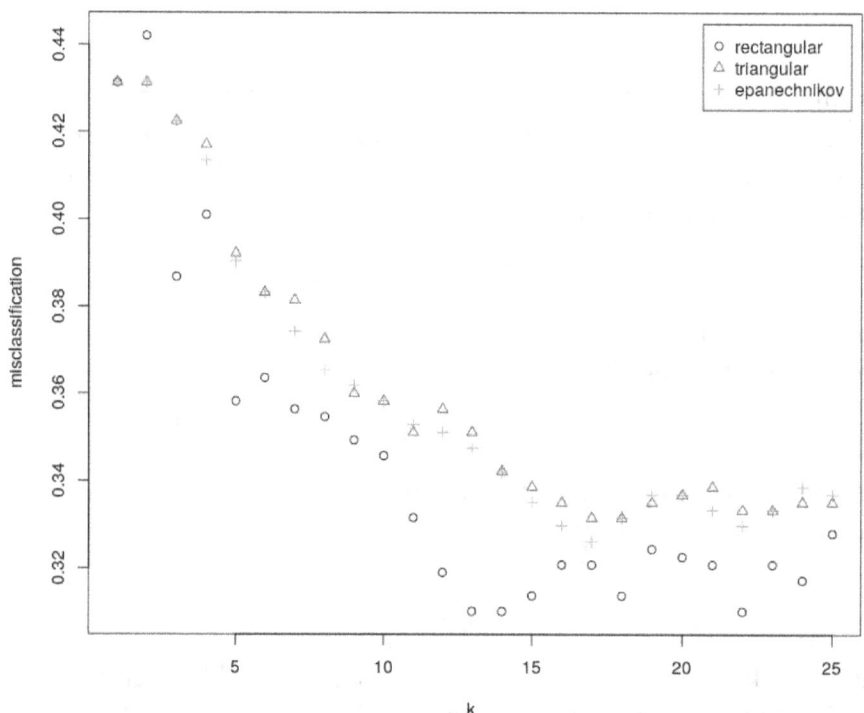

If you look at the plot you can see which value of k is the best by looking at the point that is the lowest on the graph which is right before 15. Looking at the legend it indicates that the lowest point is the "rectangular" estimate, which is the same as unweighted. This means that the best classification is unweighted with a k of 13.

Having completed this we will now turn our attention to KNN for regression.

KNN for Regression

The steps are the same for KNN regression. We will use the `Mroz` dataset again but this time we will predict income.

Step 1: Data Preparation

Below is the output for the initial code.
```
> library(Ecdat)
> data("Mroz")
> str(Mroz)
'data.frame':   753 obs. of  18 variables:
 $ work      : Factor w/ 2 levels "yes","no": 2
 $ hoursw    : int  1610 1656 1980 456 1568 2032
 $ child6    : int  1 0 1 0 1 0 0 0 0 ...
 $ child618  : int  0 2 3 3 2 0 2 0 2 2 ...
 $ agew      : int  32 30 35 34 31 54 37 54 48 3
 $ educw     : int  12 12 12 12 14 12 16 12 12 1
 $ hearnw    : num  3.35 1.39 4.55 1.1 4.59 ...
 $ wagew     : num  2.65 2.65 4.04 3.25 3.6 4.7
 $ hoursh    : int  2708 2310 3072 1920 2000 104
 $ ageh      : int  34 30 40 53 32 57 37 53 52 4
 $ educh     : int  12 9 12 10 12 11 12 8 4 12 .
 $ wageh     : num  4.03 8.44 3.58 3.54 10 ...
 $ income    : int  16310 21800 21040 7300 27300
 $ educwm    : int  12 7 12 7 12 14 14 3 7 7 ...
 $ educwf    : int  7 7 7 7 14 7 7 3 7 7 ...
 $ unemprate : num  5 11 5 5 9.5 7.5 5 5 3 5 ...
 $ city      : Factor w/ 2 levels "no","yes": 1
 $ experience: int  14 5 15 6 7 33 11 35 24 21 .
```

Here is what we need to do.

- Remove the `city` and `work` categorical variables.
- Scale the data without the dependent variable.

CHAPTER 8. K-NEAREST NEIGHBOR

- Reinsert the dependent variable into the new scaled dataset.
- Create our train and test datasets.

Below our the results for the first three bullets.

```
> Mroz$work<-NULL
> Mroz$city<-NULL
> mroz.scale<-as.data.frame(scale(Mroz[,-12]))
> mroz.scale$income<-Mroz$income
> str(mroz.scale)
'data.frame':   753 obs. of  16 variables:
 $ hoursw    : num  0.998 1.051 1.422 -0.327 0.9
 $ child6    : num  1.455 -0.454 1.455 -0.454 1.
 $ child618  : num  -1.03 0.49 1.25 1.25 0.49 ..
 $ agew      : num  -1.305 -1.553 -0.934 -1.058
 $ educw     : num  -0.126 -0.126 -0.126 -0.126
 $ hearnw    : num  0.302 -0.304 0.67 -0.394 0.6
 $ wagew     : num  0.331 0.331 0.905 0.579 0.72
 $ hoursh    : num  0.74 0.0717 1.3512 -0.5831 -
 $ ageh      : num  -1.38 -1.876 -0.635 0.978 -1
 $ educh     : num  -0.163 -1.156 -0.163 -0.825
 $ wageh     : num  -0.816 0.227 -0.922 -0.931 0
 $ educwm    : num  0.816 -0.668 0.816 -0.668 0.
 $ educwf    : num  -0.506 -0.506 -0.506 -0.506
 $ unemprate : num  -1.163 0.763 -1.163 -1.163 0
 $ experience: num  0.418 -0.698 0.541 -0.574 -0
 $ income    : int  16310 21800 21040 7300 27300
```

Here is the code for preparing our train and test set.
```
set.seed(234)
ind=sample(2,nrow(mroz.scale),replace=T,prob=c(.7,.3))
train<-mroz.scale[ind==1,]
test<-mroz.scale[ind==2,]
```

We are now ready to train our model

Step 2: Model Development

Below is the code for the cross-validation and the k recommendation from `caret`.

```
> library(caret)
> grid<-expand.grid(.k=seq(2,17,by=1))
> control<-trainControl(method="cv")
> knn.train<-train(income~.,train,method="knn",trControl=control,tuneGrid=grid)
> knn.train
k-Nearest Neighbors

561 samples
 15 predictor

No pre-processing
Resampling: Cross-Validated (10 fold)
Summary of sample sizes: 505, 505, 505, 505, 505, 505, ...
Resampling results across tuning parameters:

  k   RMSE      Rsquared   MAE
   2  9181.564  0.3482851  6761.727
   3  8836.964  0.3607600  6597.200
   4  8642.003  0.3807426  6482.830
   5  8441.377  0.4131402  6255.583
   6  8425.075  0.4171882  6230.586
   7  8392.232  0.4242485  6158.033
   8  8353.906  0.4342072  6101.823
   9  8268.204  0.4549281  6065.379
  10  8266.815  0.4588586  6020.979
  11  8283.448  0.4610113  6042.300
  12  8276.862  0.4642064  6077.392
  13  8285.443  0.4652332  6087.304
  14  8297.085  0.4699171  6072.060
  15  8256.433  0.4828859  6045.795
  16  8294.843  0.4796854  6074.407
  17  8260.320  0.4926002  6061.108

RMSE was used to select the optimal model using the smallest value.
The final value used for the model was k = 15.
```

R has recommended a k of 15. The criteria for this recommendation is the root mean square error (RMSE). The r-square is almost 50%. We can proceed to model testing.

Step 3: Model Testing

Model testing is the same as before but the assessment of the model is slightly different. We will calculate the correlation between the actual and predicted values. The higher the correlation the better.

Then we will look at two other metrics that are useful for comparison with other models but not necessarily alone. This other two metrics are the summary statistics and the mean absolute error. The summary statistics give you an idea of the range of values the model was able to predict and the mean absolute error indicates the amount of error. Both of these values mean nothing alone and are only useful if we generated other models with the purpose of comparison. Below is the code for this.

```
> library(class)
> knn.test<-knn(train[,-16],test[,-16],train[,16],k=15)
> cor(as.numeric(knn.test),test$income)
[1] 0.3415856
> summary(as.numeric(knn.test))
   Min. 1st Qu.  Median    Mean 3rd Qu.    Max.
    1.0   124.2   252.0   244.9   364.2   464.0
> mean(abs(test$income-as.numeric(knn.test)))
[1] 24101.91
```

The correlation seems somewhat low. The other two measures do not mean much without developing other models.

For our final step, we will see if weighted KNN may be more appropriate.

```
> library(kknn)
> kknn.train<-train.kknn(income~.,
+                       train, kmax = 25,distance = 2,
+                       kernel = c("rectangular","triangular", "epanechnikov"))
> plot(kknn.train)
```

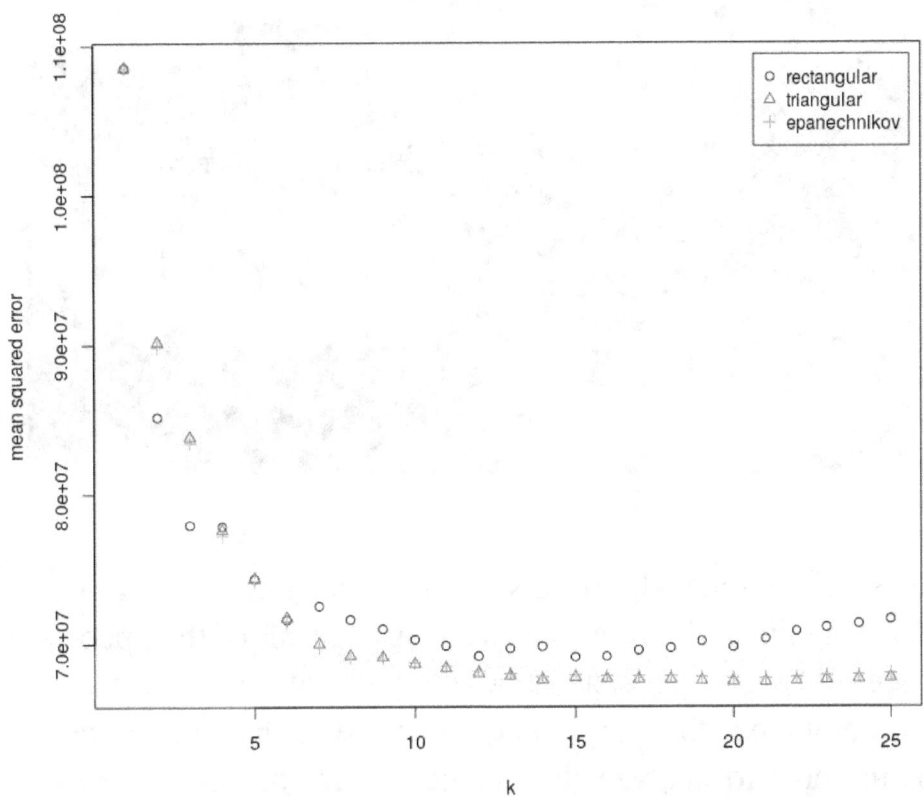

It is possible that weighted KNN would perform better but the the difference appears to be negligible.

Conclusion

KNN provides a way to extract insights from data that is highly difficult to work with due to its non-linear nature. The primary challenge is to determine the size of the "K" in the analysis. Once this is done, the algorithm should provide fairly reasonable results.

The use of complex algorithms for statistical insights has increased significantly in recent years. As a result of this, people are looking to understand how the algorithms work that are so common in the world today. In this text, Darrin Thomas explains how to analyze data using several different algorithms while using the R programming language.

SuJinSola

www.ingramcontent.com/pod-product-compliance
Lightning Source LLC
Chambersburg PA
CBHW082250220526
45469CB00009B/2949